murach's Tested Writing Methods

Mike Murach

To my 8 wonderful grandchildren with the hope that they will enjoy reading and writing as much as I do:

Dylan Stafford
Django Daillak
Hank Murach
Willa Murach
Isla Murach
Georgia Murach
Dusty Murach
Daisy Murach

murach's

Tested Writing Methods

A publisher's guide to better business writing

Mike Murach

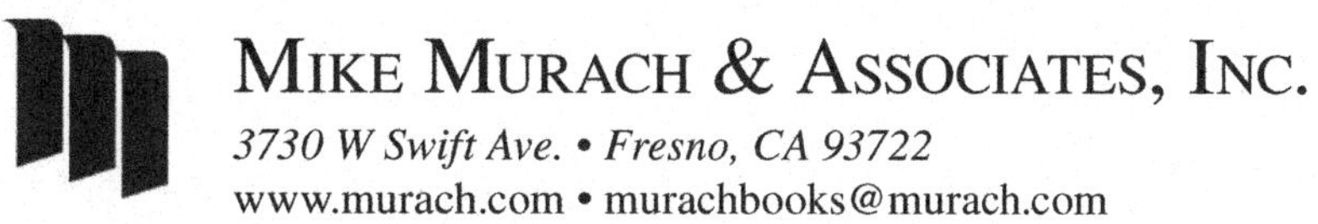
MIKE MURACH & ASSOCIATES, INC.
3730 W Swift Ave. • Fresno, CA 93722
www.murach.com • murachbooks@murach.com

Editorial team

Author: Mike Murach

Editor: Anne Boehm

Production: Juliette Baylon

Books on programming languages

Murach's Python Programming
Murach's C#
Murach's C++ Programming
Murach's Java Programming

Books on SQL and Data Analysis

Murach's Python for Data Analysis
Murach's R for Data Analysis
Murach's MySQL
Murach's SQL Server for Developers
Murach's Oracle SQL and PL/SQL for Developers

Books on web development

Murach's HTML and CSS
Murach's JavaScript and jQuery
Murach's PHP and MySQL
Murach's Java Servlets and JSP
Murach's ASP.NET Core MVC

For more on Murach books, please visit us at www.murach.com

Printed in the United States of America

10 9 8 7 6 5 4 3 2 1
ISBN: 978-1-943873-11-1

Contents

Expanded contents

Section 1 The tested writing methods

Introduction

What's unique about this book is that it presents all the writing methods in this book that you need for success on the job in just 226 pages.

What's hard to believe is that more than half of the methods aren't taught in high school courses, college courses, or in other books on writing. That goes a long way toward explaining why most business documents take too long to write, too long to read, and don't get the intended results.

The good news is that this book will help you improve your writing whether you already write well or desperately need to improve. Either way, this book will show you how to write like a pro.

Who this book is for

This book is for anyone who writes on the job and wants to learn how to write better. In fact, although the focus of this book is on business writing, the tested methods in this book work for all types of non-fiction. To write well, you need to think and write from the top-down, and that's what this book shows you how to do.

What this book does

This book presents all of the tested writing methods that you need for success on the job with the emphasis on the methods that aren't included in other books on writing. That's why the two sections in this book will help you improve in ways that other books don't even address.

Section 1 presents the tested writing methods. That includes how to use headings and subheadings to guide your readers and to plan what you're going to write…how to write paragraphs that sell your ideas...how to write sentences that are easy to read and understand...how to show the relationships between your paragraphs and sentences...how to write the first draft in record time...how to edit the first draft without "thrashing"...right down to when and how to use AI (Artificial Intelligence) writing tools like ChatGPT.

Then, section 2 shows how to use the Microsoft Word features that every writer should use: templates and styles, the outline feature, and the spelling and grammar checker. If you use Word but aren't using these features, this section will show you what you're missing. And if you aren't using Word, this section will show you why you should be.

Why you'll learn better and faster with this book

Like all our books, this one has features that you won't find in other books on writing. Here are four of them:

- This book presents a top-down approach to business writing that starts with how to plan what you're going to write and ends with how to use Word's spelling and grammar checker to catch a final error or two. No other writing book is that comprehensive.
- Above all, this book presents the writing methods that are either neglected or ignored in other writing books...especially, the methods used by professional writers.
- This book not only shows you how to write better, but also how to write faster. That starts with planning and ends with writing the first draft just once, editing it just once, and proofreading it just once. That contrasts the write, edit, rewrite, and edit again approach to writing that we call "thrashing."
- This book also shows you how to use the Microsoft Word features for writers. Why include that in a book on writing? Because most writers don't use these features, even though they will help you write faster and better.

How our "paired pages" help you learn

When you page through this book, you will see that the writing methods are presented in "paired pages" with the examples, guidelines, and procedures on the right page and the perspective and extra explanation on the left page. If at first that seems confusing, just read the text on the left page and look at the figure on the right whenever the text refers to it. By the end of the first chapter, you'll be comfortable with this presentation method.

As you will discover, this paired-pages format will help you learn faster and better because you will be reading less and studying the examples more. This is also the ideal reference format because all of the essential information is in the figures on the right pages. So, instead of digging through the text to find what you're looking for, it's all there on a single page.

What we mean by "tested" writing methods

To a large extent, writing has been treated as more of an art than a science. In business, though, you want your writing to be more science that art. You want your emails to communicate efficiently. You want your reports to draw the right conclusions and your proposals to get results. You want your marketing and web copy to sell your products and services. In short, you want your writing to get the intended results. And any writing that doesn't do that is a waste of time.

So, what we mean by "tested" writing methods is that our publishing company has tested them on the job in the real world for almost 50 years. During that time, we've used those methods to write dozens of books and to sell more than 1,000,000 copies of them. We've also used those methods to write our emails, reports, proposals, procedure manuals, web copy, and more. In short, the success of our business has always depended on how well we write...and we've used the methods in this book to do that writing.

When we started the company, though, we had to develop many of the writing methods ourselves because we couldn't find anything useful in the dozens of writing books that we reviewed. That includes: How to plan what you're going to write...how to use headings and subheadings to guide your readers...how to get the most from visuals...how to write the first draft...how to edit that first draft...and even how to write effective paragraphs.

But when it came to writing sentences, there was voluminous information in other books on writing. The trouble is that there's so much more to good writing than writing sentences. In fact, your writing can fail dismally even if your sentences are carefully crafted. So instead of focusing on all the refinements, we selected the essentials and tested them on the job. Now, chapter 4 presents those methods in a single chapter with the most important methods first.

Of course, I used our tested writing methods to write every page in this book too. That means that you'll be doing your own test of our methods as you read this book.

About the examples in this book

Many of the examples in this book are based on a report or variations of a report on eLearning that I wrote just for this book. I chose that subject not only because eLearning is becoming ever more popular, but also because how well eLearning works is still in question. That's why it's a good subject to analyze.

In fact, the general conclusion of the reports in this book is that most eLearning fails, especially for complicated skillsets. But whether or not you agree with that conclusion, keep an open mind. As you will see, the reports present plenty of information that supports that conclusion.

Also, remember that the purpose of the reports is to illustrate the use of the tested writing methods. That means that the reports aren't edited and rewritten and edited again with the goal of making them perfect. Instead, I wrote them, edited them once, and proofed them once, which is in keeping with the recommendations in chapter 6.

Of course, this book also presents many examples taken from real-world emails, white papers, reports, proposals, and books. But most of those examples are the *before* examples that this book shows how to improve. That's because it's hard to find examples in the real world that we could use as the *after* examples.

In fact, I've analyzed hundreds of business documents during the last 50 years, and I think it's fair to conclude that most business documents don't get the intended results. In other words, most business writing fails. That's why learning how to use the tested writing methods will help ensure your success on the job.

About the download for this book

In appendix A, you can learn how to download the reports and templates for this book and install them on your system. The download includes both Word and RTF versions of two of the eLearning reports that are used to illustrate the methods in this book. After you download the reports, you can analyze and experiment with them.

The download also includes the two Word templates for reports that are presented in chapter 8. After you download them, you can see how easy it is to start new documents from Word templates that contain all the styles, formatting, and starting content that you need. You can also see how easy it is to modify the styles in a new document and how easy it is to modify a template.

Please let us know how this book works for you

When I started this project, I had two goals. First, I wanted to present all the writing methods that you need for success on the job with special emphasis on the methods that aren't included in other books on writing. Second, I wanted to present those methods in a new top-down way that makes those methods easier to master.

Now, if this book helps you write better and faster, I would love to hear from you. If this book helps you enjoy writing more than ever, I would like to hear that too. And if you have any suggestions for how I can make this book better, please let me know.

Thanks for buying this book. And all the best with your writing.

Mike Murach, Author and Founder
Mike Murach & Associates, Inc.
mike@murach.com

Section 1

The tested writing methods

In this section, you'll learn how to use *all* of the tested writing methods that you need for successful business writing in just 156 pages! That includes how to use headings and subheadings to plan what you're going to write... how to write paragraphs that sell your ideas...how to write sentences that are easy to read and understand...when and how to use visuals...how to write and edit without "thrashing"...right on down to when and how to use AI writing tools like ChatGPT.

Because each of these chapters is chockful of information, you won't be able to remember and apply all the methods in a chapter after just one reading. But as you will see, each chapter is designed so it works first as a tutorial, and then as an on-the-job reference. Just be sure that you eventually read all of the chapters in this section because each one presents methods that will help you improve your writing.

How to use headings to plan short documents and guide your readers

To get you started with the tested writing methods, this short chapter shows you how to use headings in short documents like emails and reports that are from one to three pages long. If you don't use headings in short documents like that, you'll find out why you should. And if you do use headings, you'll learn how to use them so they do the best job of guiding your readers.

How to use headings in short documents

Let's agree that a *short document* is a document like an email or short report that's from one to three, single-spaced pages. So with an average of about 8 paragraphs per page, a short document will require from 8 to 24 paragraphs. Then, the question that this chapter answers is: How do you plan the contents for a short document?

As you will see, the answer is to divide the document into topics that will require from one to six paragraphs each and to use headings to identify each of those topics. Then, the headings will not only guide your readers through the document, but also make it easier for you to write the document.

Why you should use headings in your documents

The example in figure 1-1 shows the first six paragraphs of a document that's nine paragraphs long with no headings. Now, be honest. Is that a document you would like to read? Even for a lifetime reader like me, documents like that look daunting. And all too often they turn out to be the way they look: deadly dull.

In fact, most people today are so busy that they don't want to take the time to read a long document. They've even created an acronym for that: *TL:DR* (*Too Long: Didn't Read*). Yes, it's hard to imagine that some of your readers may not want to take the time to read 12 or 14 of your carefully-crafted paragraphs, but that's the way the world is.

So the first reason for using headings in your documents is to combat the *TL;DR syndrome*. That way, your readers can at least find and read the topics that they are interested in. But the headings will also make it easier for your readers to follow the flow of your topics and ideas.

The good news is that using headings will also make it easier for you to write the documents. That's because you won't have to provide the transitions that the headings make obvious. You'll see how this works in the chapter on writing paragraphs.

But be aware right now that there's a right way and a wrong way to use headings. You don't just add them in after you've written the document. And you don't use general headings like Introduction, Four Reasons, and Conclusion. Instead, you plan the headings before you write one word of the text.

The start of a short report that doesn't have headings

Why most eLearning courses fail

Okay, we accept that eLearning is a large market that is growing every day. But shouldn't someone be asking whether these courses deliver the results that they promise? And even if they do deliver the results, shouldn't someone be asking whether eLearning is the most efficient way to deliver those results? What follows then are some of the reasons why most eLearning courses fail.

Let's start with the limitations of video since video is the primary medium for delivering the content in many eLearning courses. First, let's agree that video is effective whenever you can learn more by seeing how something is done than by reading about it, like playing tennis or dancing. For most courses, though, video has some serious limitations.

The first problem is that most people can read from 3 to 5 times faster than a video presents information. So even if a video is effective, it's an inefficient way to learn. Besides that, it's hard to skip over information that you already know, and that slows you down even more.

With video, it's also hard to refer to the information that you need to review. But that's essential for any subject for which it's impossible to remember all of the details. For example, when you watch a video of a computer procedure, like how to use an IDE, it's difficult to remember all the steps when you try to do them yourself. It's also hard to do the steps at the same time that you're watching the video because you have to switch back and forth between the video and the software product that you're trying to master. This is a major problem for any difficult subject.

Slides are the other primary medium for delivering eLearning content. In fact, slides have been commonly used for classroom instruction for many years. In that scenario, the instructor goes through one slide at a time and explains or enhances its content. And now, slides are a primary medium for delivering the content in eLearning courses. The trouble is that slides also have some serious limitations, especially for difficult subjects.

The primary criticism is that slides force you to break down the content into chunks that are so small that it's hard for the student to see how the chunks relate to each other. But it's seeing those relationships that lead to deeper understanding and the ability to apply what you've learned. For programming subjects, this means that there's no way to present something like a lengthy program listing without breaking it down into chunks that don't make sense by themselves.

The other criticism of slides is that they force you to present the content in a rigidly sequential way. Then, to see how the chunks are related, the student must go forward

How the proper use of headings will improve your writing

- They will help combat the *TL:DR (Too Long; Didn't Read) syndrome.*
- They will make it easier for your readers to understand and review your documents.
- They will make it easier for you to write your documents.

Description

- Whenever you write an email or report that's going to be much longer than 5 or 6 paragraphs, you should plan the headings before you start writing.

Figure 1-1 Why you should use headings in your documents

Why you should use a heading plan to plan the headings

Of course, most writers and publishers agree that headings are a good thing. That's why most articles, reports, proposals, and books have them. But if that's true, shouldn't you plan the headings before you start writing so they do a good job of guiding your readers? In fact, if you try that, you'll find that planning the headings is also the best way to plan what you're going to write.

So in figure 1-2, you can see how a *heading plan* can be used to plan the headings for a document. Quite simply, the heading plan shows the *headings* that the document will use. And these headings identify the *topics* that the document will present as well as the sequence in which the document will present them.

In the first example in this figure, you can see four headings that you could use for the document in figure 1-1. Those headings divide the document into four topics. The first three will present specific types of eLearning problems. The last one will provide the author's recommendation. You can imagine how much easier it would be to understand and review the document if it had used those headings.

In the second example, you can see a heading plan for a one-page email that requires just two headings. This just shows that headings can be used whenever a document can be divided into two or more topics. And that's true for most documents that require more than 5 or 6 paragraphs.

Once you create the heading plan for a document, you can write the paragraphs that provide the content for each heading. But that's relatively easy to do because most topics will require from one to six paragraphs. In short, you've broken down the content of the document into manageable units of writing.

Example 1: The heading plan for "Why most eLearning courses fail"

The limitations of video
The limitations of slides
What most eLearning courses don't include
My recommendation

A heading that will help guide the readers

> What follows are some of the reasons why most eLearning is both ineffective and inefficient. These reasons apply to most eLearning courses. And they are especially true for programming courses, which of course is the market that we're in.
>
> **The limitations of video**
>
> As I have mentioned, video is the primary medium for delivering the content of many eLearning courses. And there's no question that video is effective whenever you can learn more by seeing how something is done than by reading about it. For instance, video is effective for showing how to dance, how to play tennis, and how to use the IDE (Integrated Development Environment) for a programming course.
>
> But for most of the content of a course, video has some serious limitations. Here are some of

Example 2: The heading plan for an email on web problems

Our eBook delivery system has an occasional bug
Some of our marketing copy needs to be updated

What a heading plan is

- A *heading plan* for a short document consists of the *headings* that identify the *topics* that the document will present in the sequence in which they will be presented.

How your headings should guide your readers

- The headings should clearly identify the topics in a document. That will not only help your readers see the structure of the document but also help them find and review topics later on.

The two mistakes that many business writers make

- Not using headings at all in their short documents.
- Using headings that don't identify the topics in the document.

Description

- Even short documents of from one to three pages (or from 8 to 24 paragraphs) should be divided into writing units called *topics*. And each topic should be identified by a *heading*.
- To plan the headings that will be used when you write the document, we recommend the use of a *heading plan*.

Figure 1-2 Why you should use a heading plan to plan the headings

A three-step procedure for planning the headings

Figure 1-3 presents a three-step procedure for planning the headings for short documents. In most cases, that won't take long because your heading plan will consist of just three or four headings. But if it's an important document, even if it requires just a few headings, it makes sense to use the three-step procedure in this figure to get the headings right.

In step 1, you list the topics that you think should be in the document. Then, you select the topics for the document based on who your audience is and what your purpose is.

In step 2, you arrange the topics in the best sequence for presentation. For most short documents, this means that you're arranging from three to five topics, which means that you don't have that many options. In this example, the writer has made a few adjustments to the starting sequence so the topics move from what he thinks are the most important to the least important.

At this point, the topic names are likely to be in a language that makes sense to you, the writer. Then, in step 3, you edit or rewrite those topic names so they become headings that clearly identify the topics for the readers. To do that, you can use the two guidelines in this figure.

Above all, you need to be specific so your readers can easily tell what each topic is going to present. But it also helps to use consistent language structures. For instance, "The limitations of video" and "The limitations of slides" have a parallel structure that works better than "The limitations of video" and "Slide problems" because the parallel structure implies that the topics will have similar approaches to the content.

As you get experience with this procedure, you'll find that you can merge all three phases. So right from the start, you will list the topics for the document in language that is appropriate for headings, and you will sequence the headings as you list them.

How do you implement a heading plan when you're using a word processor? The easiest way is to start a new document, enter the headings into it, and apply heading styles to the heading paragraphs. Then, you can enter the paragraphs that present the content for each heading.

If you're using Microsoft Word, a better alternative is to use its outline feature to enter the headings into a new document. Then, when you've got the headings the way you want them, you can switch from Outline view back to Print Layout view and start writing. You can learn more about this in chapter 9.

If you're writing an email, though, your email application probably doesn't have headings styles. In that case, you can just put the headings in separate paragraphs and boldface them so the reader can tell what they are. That's worth doing because few people today want to read a long email that doesn't have any headings.

How to create a heading plan for a short document

1. List and select the topics for the document.
2. Arrange the topics in the best sequence for presentation.
3. Edit and improve the topic names so they work as headings.

How this works for a short report: "Why most eLearning fails"

1. **List and select the topics for the document**

   Slide issues
   Video issues
   The focus on technology
   What's missing

2. **Arrange the topics in the best sequence for presentation**

   Video issues
   Slide issues
   What's missing
   The focus on technology

3. **Edit and improve the topic names so they work as headings**

   The limitations of video
   The limitations of slides
   What most eLearning courses don't include
   Why the focus on technology is counterproductive

The primary guidelines for writing headings

- Be specific so the reader can easily tell what each topic is going to present.
- Use consistent language structures.

Two ways to implement the heading plan for a short document

- Start the document, enter the headings as separate paragraphs, and apply the heading style to each paragraph.
- Use the outline feature of Microsoft Word to enter the headings. When you're done, you can switch from Outline view to Print Layout view and start writing.

Description

- As you get experience with the thought process that's outlined above, you'll merge the three phases into one in which you list the headings that you're going to use in the best sequence for presentation and with the best phrasing right from the start.

Figure 1-3 A three-step procedure for planning the headings for a short document

Related skills

At this point, you should be able to develop heading plans of your own. But to give you more perspective on that, here are some related skills.

How to choose the organization for a document

For many short documents, the presentation sequence is obvious. But as a document gets longer or more complex, you may have to think about the best way to present its information. That's why figure 1-4 presents some ideas that may help you organize a document.

At the top of this figure, you can see some of the common organizational methods for business documents. This list starts with the *order of importance*, also known as the *order of decreasing importance.* With this organization, you just present the topics from the most important to the least important. The first example in this figure provides a general heading plan for a document that's organized that way.

Another useful way to organize a document is from *simple-to-complex*. In this case, you start by introducing the subject. Then, you build on that with successive layers of information. This is illustrated by the second example in this figure, which is for the topics in an eLearning report. Here, the readers are introduced to eLearning before they learn about some typical eLearning courses. Then, they learn about some of the shortcomings of eLearning, after which they consider the ways to fix those shortcomings.

The other organizational methods are based on procedures, functions, time, or geography. If you're writing documents that require one of these organizations, you'll find that they're relatively easy to implement. For instance, the third heading plan in this figure has procedural organization, and the fourth has geographical organization.

In the real world, however, the organization for a document doesn't always fall into one of these categories. Instead, most documents require some combination of simple-to-complex and order of importance. So when in doubt, let those two organizational methods guide your thinking, but don't try to force your content into either one of them. In the end, you have to decide what will work best for your audience and purpose.

Organizational methods for business documents

- Order of importance, also known as *order of decreasing importance*
- Simple-to-complex
- Procedural or functional
- Chronological or geographical

A heading plan that's organized by order of importance

Major findings
Minor findings
Miscellaneous findings

A heading plan for an eLearning report that goes from simple to complex

Introduction to eLearning
Typical eLearning courses
Why most eLearning fails
How our eLearning courses could succeed

A heading plan that has procedural organization

Step 1: Select and sequence the topics
Step 2: Add and sequence the subtopics
Step 3: Convert the topic and subtopic names to headings and subheadings
Step 4: Analyze and improve the heading plan

A heading plan for a marketing report that has geographical organization

Good news from Minnesota and Michigan
Problems in Wisconsin
Continued progress in Illinois and Indiana

Description

- In practice, the content for many business documents doesn't fit into one of these organizations. Instead, the heading plans need to be based on some combination of simple-to-complex and order of importance.
- What's most important is that you take the time to think carefully and critically about the sequence of the topics that you're going to present.

Figure 1-4 How to choose the organization for a document

How to improve your headings and document titles

Because headings are so important, figure 1-5 presents four more guidelines for writing them. The first three also apply to the titles that you use for your documents.

The first guideline is to write headings that "promise a benefit." To some extent, all headings promise the benefit of useful information. But you can improve those headings by being more specific about how useful that information is going to be. This is illustrated by the headings in the first group of examples. For instance, "Why eLearning matters" promises more of a benefit than "Introduction." And "How to open a new account" promises more of a benefit than "New accounts."

Often, all you need to do to promise a benefit is to be more specific about what the reader is going to learn. To that end, words like *when, why,* and *how* can make a benefit more obvious. And headings that promise a number of items of information, like "Six ways to organize a document," provoke curiosity in a way that other wording can't. Note, for example, how much better the replacements are for the "Introduction" and "Conclusion" headings.

The second guideline is to use long headings whenever they're appropriate. This is illustrated by the second example in this figure. These were taken from a newspaper article, and they make it easy for the readers to read about the stock market changes that they're most interested in. Yes, these headings could be shorter, but they're effective just the way they are.

The third guideline is to avoid the use of *verbals* (or "ing" words) like *using* and *understanding* in your headings. This is illustrated by the third group of examples. The reason for avoiding these words is that they lead to headings that aren't as specific as they ought to be.

The last guideline in this figure is to capitalize just the first word in each heading plus any required capitalization (called *first cap*). That may sound trivial but lowercase headings are easier to read than headings that capitalize the first letter of each major word (called *title case*), and they're much easier to read than headings that are in all capital letters (called *all caps*). So make it a rule: Don't ever use all caps for your headings!

All of these guidelines also apply to document titles. This is illustrated by the last group of examples. As you finish up the heading plan for a document, you know what it's going to present, so that's a good time to write the title for the document. Then, if your title promises a benefit, your document will be off to a great start.

Four more guidelines for writing headings

- Whenever appropriate, promise a benefit (and maybe provoke curiosity).
- Don't be afraid to use long headings.
- Avoid verbals like *using, understanding, exploring,* and *comparing.*
- Only capitalize the first letter in each heading (plus any required capitalization).

How promising a benefit can improve your headings

Original	**Promising a benefit**
Introduction	Why eLearning matters
New accounts	How to open a new account
Organizational methods	Six ways to organize a document
Conclusion	Why we should start a pilot program

Long headings in an article on the stock market: "U.S. sees major shift over 10-year bull market"

Index funds have become the default way to invest

Investing has become more accessible and affordable

The industry now caters to investors who don't want to figure out how to invest their nest eggs

What hasn't changed is that stock investing is still risky

How rewriting verbals can improve headings

Original	**Improved**
Using eLearning products	How to use eLearning products
Comparing the authoring tools	How we rate the authoring tools

How promising a benefit can improve your document titles

Original	**Improved**
Collection policies	Three recommendations for how we can improve our collection policies
The eLearning market	Why we should enter the eLearning market

Description

- The primary benefit that readers want to get from any document is useful information. That's why report titles and headings that "promise a benefit" can make your documents more appealing...and if they provoke curiosity, that's good too.
- Long headings work well because they can be more specific. And headings that don't use *verbals* like "using" and "comparing" tend to be more specific.

Figure 1-5 How to improve your headings and document titles

Perspective

As you have just seen, it's relatively easy to plan the headings for short documents like emails and short reports. And yet, all too many business writers don't use headings at all...or use headings that don't guide the reader. If you're one of those writers, using headings the right way is an easy way to improve your short documents.

If you also write long documents, be sure to read the next chapter too. There, you'll learn how to plan both the headings and the subheadings for a document. As you will see, that's a bit more difficult, but it's critical to the success of documents like reports and proposals.

Terms

heading plan
topic
heading
TL:DR syndrome
order of importance
order of decreasing importance
simple-to-complex
verbal
first cap
title case
all caps

Summary

- In general, you should use headings in any document of more than 5 or 6 paragraphs. That will help combat the *TL:DR syndrome*. That will make it easier for your readers to understand what you've written. And that will make it easier for you to write the text.
- A *heading plan* consists of the *headings* that identify the *topics* that the document will present. It also shows the sequence in which the topics will be presented.
- To create a heading plan for a short document, you (1) list and select the topics, (2) sequence them, and (3) edit or rewrite the topic names so they work as headings.
- When you edit or rewrite topic names as headings, you need to be specific so the reader can tell what each topic is going to present. You should also use consistent wording and language structures.
- Two common organizations for business documents are *order of importance* (or *order of decreasing importance*) and *simple-to-complex*. In practice, though, most business documents don't fall into a single organizational category.
- To improve your headings, you can promise a benefit, use long headings and subheadings, and avoid *verbals* like *using* and *understanding*.
- When you use *first cap* for your headings, instead of *title case* or *all caps*, you make a minor but professional improvement to your headings.
- The guidelines for writing headings also apply to document titles, except that title case is commonly used instead of first cap.

2

How to use headings and subheadings to plan long documents

In the last chapter, you learned how to use headings to plan short documents and guide your readers. Now, in this chapter, you'll learn how to use both headings and subheadings to plan long documents like reports and proposals. Those headings and subheadings will also guide your readers and help them find and review specific topics and subtopics.

If you write more than a few long documents a year, you're going to find that this chapter will be one of the most valuable chapters in this book. That's because most business writers don't have an effective way to plan what they're going to write, and poor planning usually leads to poor writing. Curiously, though, most books on writing treat planning lightly, if at all. In fact, headings aren't even mentioned in many of those books.

An introduction to the use of headings and subheadings

The focus of this chapter is on how to use headings and subheadings to plan long documents like reports and proposals. But before you learn that, you need to understand how headings and subheadings should be used to guide your readers.

How headings and subheadings should guide your readers

Today, people have access to more information than ever before. The trouble is that they don't have time to read everything.

One way to help them is to provide headings and subheadings that identify the topics and subtopics that make up a document. That will guide the readers through the content of the document. And that will make it easy for them to refer back to those topics and subtopics after they read the document.

This is illustrated in figure 2-1. In the first example, you can see a heading and a subheading in the body of a report on eLearning. This shows how headings and subheadings should be used to identify the topics and subtopics within a document.

In the second example, you can see how a report on eLearning can be divided into headings and subheadings that guide the readers through the content. In this example and in the other examples in this chapter, the subheadings are indented so you can tell them from the headings.

In the third example, you can see how a heads-up report on web problems can be divided into headings and subheadings. Because the subheadings clearly identify the problems that need to be fixed, the resulting document will be a good reference for making sure that the problems get fixed.

Now, imagine these documents if they didn't use headings and subheadings. Even for a lifetime reader like me, the acronym *TL;DR (Too Long; Didn't Read)* comes to mind. Because whenever I see long sequences of paragraphs, I wonder whether I'm going to have to plow my way through a tedious argument instead of getting useful information and perspectives. And all too often, it's the tedious argument.

For the Millennial and Gen Z generations, of course, TL;DR is their default position. If you're a writer, that's maddening. But that's why it's so important to use headings and subheadings that guide your readers.

How headings and subheadings should guide the readers

Overall, colleges seem to be satisfied with the eLearning courses that they use. And why wouldn't they be? They don't have to develop the courses themselves. The courses are easier to run than traditional courses. And the costs are manageable for the students. *The trouble is that most eLearning courses fail.*

Why most eLearning courses fail

Okay, we accept that eLearning is a large market that is growing every day. But shouldn't someone be asking whether these courses deliver the results that they promise? And even if they do deliver the results, shouldn't someone be asking whether eLearning is the most efficient way to deliver those results? What follows then are some of the reasons why most eLearning courses fail.

The limitations of video

Video is the primary medium for delivering the content of many eLearning courses. And there's no question that video is effective whenever you can learn more by seeing how something is done than by reading about it. For instance, video is effective for

The headings and subheadings for an eLearning analysis

Why most eLearning courses fail
- The limitations of video
- The limitations of slides
- The missing ingredients

Why our eLearning courses would succeed
- Our proven books will be the primary delivery system
- Our eLearning components will add to the effectiveness of our books

My recommendation

The headings and subheadings for a report on web problems

Our eBook delivery system has occasional bugs
- Some of our confirming emails are treated as spam
- Some of our Redemptions Codes don't work the first time

Some of our marketing copy needs to be updated
- The "About our eBooks" page
- The "Courseware for Instructors" page
- The "How to get your eBook from VitalSource" page

How your headings and subheadings should guide your readers

- The headings and subheadings should clearly identify the topics and subtopics in a document. That will not only guide your readers through the document but also make it easy for your readers to refer to specific topics and subtopics later on.
- The headings and subheadings should also show the structure of the content. That will make it easier for your readers to see the relationships between the topics and subtopics.

Description

- When you write long documents like reports and proposals, you usually need to use both headings and subheadings to identify the topics and subtopics in the document.

Figure 2-1 How headings and subheadings should guide your readers

Why you should use a heading plan to plan long documents

Of course, most writers and publishers agree that headings are a good thing. That's why most articles, reports, proposals, and books have them. But if that's true, shouldn't you plan the headings and subheadings before you start writing so they do a good job of guiding your readers? In fact, if you try that, you'll find that planning the headings and subheadings for your readers is also the best way to plan what you're going to write.

So in figure 2-2, you can see how a *heading plan* can be used to plan the contents of a document. Quite simply, the heading plan shows the *headings* and *subheadings* that the document will use. And these headings and subheadings identify the *topics* and *subtopics* that the document will present as well as the sequence in which the document will present them.

In the example in this figure, you can see the headings and subheadings for a lengthy report on eLearning. This shows that headings and subheadings are a lot more important in a long document than a short one. That's why it's so important to do an effective job of planning them. And creating a heading plan is the best way to do that.

Once you create a heading plan, you just need to write the paragraphs that provide the content for the headings and subheadings. That's because the heading plan represents the natural structure and sequence of a written document. That's why this is such an effective way to plan what you're going to write.

The heading plan for an analysis of the eLearning market

Introduction to eLearning
    What eLearning is
    The primary components of eLearning
    eLearning and the Learning Management System (LMS)
A quick tour of some typical eLearning courses
    MOOCs
    Commercial video courses
    Corporate eLearning courses
    College eLearning courses
Why most eLearning courses fail
    The limitations of video
    The limitations of slides
    The missing ingredients
Why our eLearning courses will succeed
    Our proven books will be the primary delivery system
    Our eLearning components will add to the effectiveness of our books
My recommendation

What a heading plan is

- A *heading plan* consists of the *headings* and *subheadings* that identify the *topics* and *subtopics* that the document will present.

Description

- If headings and subheadings are one of the keys to successful business writing, it makes sense to plan the headings and subheadings before you start writing.
- To plan the headings and subheadings for a document, we recommend the use of a *heading plan*. To distinguish the subheadings from the headings in a heading plan, you indent the subheadings.
- Because a heading plan represents the natural structure and sequence of a written document, it also works as a plan for writing the document. To do that, you just write the paragraphs that provide the content for each heading and subheading.

Figure 2-2 Why you should use a heading plan to plan long documents

How to use a heading plan to plan long documents

Now, that you know what a heading plan is, you're going to learn a four-step, top-down procedure for creating one. As you will see, that will help you divide a large writing project into manageable pieces. That of course is what you need to do for any complex project. Then, instead of writing one long sequence of paragraphs, you write the paragraphs for one topic or subtopic at a time. That's something that any writer should be able to do, and you end up with a document that will also work better for your readers.

Step 1: Select and sequence the topics

The first planning step is to select and sequence the headings for the document. This is similar to what you do for short documents. But since the heading plan is going to be longer and more complicated, it makes sense to use the outline feature of a word processor like the one for Microsoft Word. As chapter 9 shows, that makes it easy to change headings to subheadings, re-sequence headings and subheadings, and more.

This is illustrated in figure 2-3. Here, the first example shows a Word outline that the writer has used to organize the research for a report. Then, the second example is a Word outline that shows the start of a heading plan for the report. At this point, the headings have been selected and sequenced based on the audience and purpose for the report.

Of course, you need to have a clear view of who that audience is and what your purpose is as you do this step. But for most business writing, that isn't an issue. You're writing a report for your boss, a proposal for a customer, or a memo for your staff. There's usually no mystery to it. Yes, you need to keep your audience and purpose in mind as you plan and write, but otherwise, get on with it.

Please note in these examples that the organization of the research and the heading plan for the topics have two different structures. That makes sense because you normally organize your research based on the structure of the subject or content. In contrast, a heading plan should focus on which topics need to be included in the document and what the best sequence for presenting those topics is.

In fact, one of the weaknesses of many business documents is that they present the information with an organizational or hierarchical structure instead of a presentation structure. That's especially true for college textbooks and academic writing, and you'll see an example of that near the end of this chapter.

The start of a Word outline that organizes the research for a report called "Is eLearning an opportunity for us?"

- eLearning components
  - slides
  - videos
  - quizzes
  - animations
  - simulations
  - interactions
- eLearning authoring tools
  - PowerPoint add-ons
    - iSpring
  - Standalone tools
    - Articulate
    - Captivate
- Typical eLearning courses
  - MOOCs
  - Commercial video courses
  - Corporate eLearning courses
  - College eLearning courses
- LMS functions
  - Register students
  - Deliver eLearning modules
  - Implement group communication
  - Track student performance
  - Get eLearning module data
- LMS implementation
  - The importance of SCORM
  - The 4 levels of SCORM compliance

The Word outline for a heading plan after the topics have been selected and sequenced

- Introduction to eLearning
- The eLearning components
- Typical eLearning courses
- Why most eLearning fails
- How our eLearning courses could succeed
- Conclusion

Description

- In this step, you select, and sequence the topics for the document based on your audience and purpose. Your goal is to include all the content that will help the document be successful, but to delete or omit any unnecessary content.
- This is an important step because two common failings in business writing are (1) not presenting all of the essential information, and (2) including information that doesn't contribute to the purpose of the document.

Figure 2-3 Step 1: Select and sequence the topics

Step 2: Add and sequence the subtopics

After you select and sequence the topics for a document, you need to add the subtopics that make up the topics. This is illustrated in figure 2-4. Here, each of the first three topics has three subtopics, and the fourth topic has two.

After you add the subtopics for each topic, you try to put them in the best sequence for presentation. For instance, you might ask whether "Slide courses" should come before or after "Video courses." Or whether "Slide criticisms" should come before "Video criticisms."

As you add the subtopics, you may decide that the topics aren't quite right for what you're trying to do. Then, you can adjust the topics as needed. For instance, the topic on "eLearning components" in the original topics list in this figure has become a subtopic under the "Introduction to eLearning" topic in the expanded heading plan.

As you develop a heading plan, you may get the feeling that you're going to forget an idea or two that you want to include in a topic or subtopic. In that case, if you're using the outline feature of Microsoft Word, you can add a third level of indentation that summarizes those ideas below the topics or subtopics that they relate to. Then, you can hide that level when you want to view or print just the topics and subtopics.

This is illustrated by the third example in this figure. Here, a third level has been added to the "Video criticisms" subtopic. This level lists the ideas for the paragraphs that will be written for that subtopic. In most cases, you won't need to plan down to that level, but sometimes it puts your mind at ease. You'll learn more about that in the next chapter.

When you're through with your first draft of the heading plan, you should be confident that it presents all the information that the document requires, with little or no information that isn't required. You should also be confident that the topics and subtopics are in an effective sequence for presentation.

Keep in mind, however, that some documents may not require both headings and subheadings. Instead, a sequence of four or more headings may be all that the document requires. If that's the case, you'll probably realize that as you do step 1. If not, you will for sure discover that when you try to do this step. Then, you can go back to the procedure shown in chapter 1.

The Word outline for the topics in a heading plan

- Introduction to eLearning
- The eLearning components
- Typical eLearning courses
- Why most eLearning fails
- How our eLearning courses could succeed
- Conclusion

The Word outline after subtopics have been added to the heading plan

- Introduction to eLearning
  - The definition
  - The eLearning components
  - The Learning Management System (LMS)
- Typical eLearning courses
  - MOOCs
  - Video courses
  - Slide courses
- Why most eLearning fails
  - Video criticisms
  - Slide criticisms
  - Missing components
- How our eLearning courses could succeed
  - Our books as the primary medium
  - Plus strategic eLearning components
- Conclusion

The procedure for adding the subtopics

1. For each topic, add the subtopics that make up the content for the topic.
2. Arrange the subtopics for each topic into the best presentation sequence. As you do that, you may decide to combine, delete, or re-sequence some of the topics.

How you can use a third level to plan the paragraphs for a topic

- Why most eLearning fails
  - Video criticisms
    - The details are hard to remember
    - It's hard to review the details
    - It's hard to skip what you already know
    - Reading is faster
  - Slide criticisms
  - Missing components

Description

- In general, you shouldn't use more than two levels of topics and subtopics because that makes the structure too complex for an effective presentation.
- It sometimes makes sense to plan the paragraphs for a subtopic or two while they're fresh in your mind. To do that, you can take the heading plan down one more level.

Figure 2-4 Step 2: Add and sequence the subtopics

Step 3: Convert the topic and subtopic names to headings and subheadings

Once you've got the topics and subtopics the way you want them, you should convert the topic and subtopic names to the headings and subheadings that you're going to use when you write the document. In figure 2-5, for example, you can see the heading plan in figure 2-4 after its names have been converted to headings and subheadings. In this step, you're trying to change the language in the topics and subtopics so it has more meaning for the readers.

After the example, this figure presents some guidelines for writing headings and subheadings. These are the same as the ones in chapter 1. Here again, what's most important is to be specific so the reader can tell what each topic and subtopic is going to present. But it's also important to use consistent language structures because that makes it easier for the reader to see the relationships between the headings as well as the relationships between the subheadings that are used for each heading.

As you rewrite the topic and subtopic names, you may realize that the sequence or structure of the topics and subtopics can be improved. That's because your focus on the language that's best for your readers may put a different perspective on what you're going to present. In that case, you just adjust the heading plan and continue.

This is also a good time to reflect on the types of organization that you're using for the topics in the document as well as the subtopics in each group. As you learned in chapter 1, the two organizations that you should keep in the back of your mind are order of decreasing importance and simple-to-complex. But note that the topics for the document may have one organization, and the subtopics within a topic may have another organization.

For instance, you could say that the topics in the heading plan in this figure move from the simple to the complex. That's also true for the subtopics in the "Introduction to eLearning." But that isn't true for the subtopics that make up the other topics. In fact, they aren't in any defined organization. Instead, they're in the sequence that the author thinks will work the best for the readers.

The Word outline after the topics and subtopic names have been converted to headings and subheadings

- Introduction to eLearning
  - What eLearning is
  - The primary components of eLearning
  - The Learning Management System (LMS)
- Typical eLearning courses
  - MOOCs
  - Commercial video courses
  - Corporate eLearning courses
  - College eLearning courses
- Why most eLearning fails
  - The limitations of video
  - The limitations of slides
  - The missing ingredients
- Why our eLearning courses would succeed
  - Our books would be the primary delivery system
  - The eLearning components would enhance the effectiveness of our books
- My recommendation

Guidelines for writing headings and subheadings

- Be specific so the reader can tell what each topic and subtopic is going to present.
- Use consistent language structures.
- Whenever appropriate, promise a benefit (and maybe provoke curiosity).
- Don't be afraid to use long headings and subheadings.
- Avoid verbals like *using, understanding, exploring,* and *comparing.*
- Only capitalize the first letter in each heading and subheading (plus any required capitalization).

Description

- In this step, you convert the topic and subtopic names to the headings and subheadings that will be used in the document. To do that, you convert the language that has meaning for you to the language that will mean the most to your readers.
- As you convert the topic and subtopic names to headings and subheadings, you will often make some improvements to the structure of the heading plan.
- Although it sounds trivial, capitalizing just the first letter of your headings and subheadings makes them easier to read. This is called *first cap*, and this contrasts *title case* (the first letters in all major words are capitalized) and *all caps* (all capital letters).

Figure 2-5 Step 3: Convert the topic and subtopic names to headings and subheadings

Step 4: Analyze and improve the heading plan

After you complete the first draft of a heading plan, you should review and analyze it, especially if you're writing an important document. To help you do that, figure 2-6 presents a reviewing checklist. Most important is to make sure that the heading plan includes all of the information that the document should present.

When you're sure that it does, you can focus first on the sequence of the headings and then on the sequence of the subheadings for each heading. To do that, you can use Word's outline feature to hide one or more levels of a heading plan, as shown in chapter 9. In the first example in this figure, just the headings are displayed, which makes it easy to check whether those headings are in the right sequence and have consistent wording.

In the second example, the subheadings for just one heading are displayed. That makes it easy to check whether those subheadings include all of the subtopics that are related to that heading, whether the subheadings are in the right sequence, and whether they have consistent wording.

You should also make sure that each subtopic is *logically subordinate* to the heading that it is beneath. This means that the subtopic is contained within the context of the topic. For instance, "commercial video courses" are contained within the subject of "typical eLearning courses," but "eLearning authoring tools" aren't. This is just common sense, but a common fault is to slip in a subtopic that has nothing to do with the topic that it is beneath. And that can only confuse the reader.

One other factor to consider is the *span of control* for each heading. That refers to the number of subheadings that the heading has. When a heading has more than 7 subheadings, for example, a reader's mind starts to boggle. For that reason, you should try to hold the maximum span of control to 7, with 2 through 5 as the optimum span of control.

Much less of a problem, though, is the minimum span of control. Although a table of contents looks better when the minimum number of subheadings is 2, a document will read well even if it has some headings that have no subheadings.

After you do your analysis, you can make the improvements to the heading plan that your analysis suggests. When you're done, you should be able to look at your heading plan and feel confident that it's a great start for a successful document. So all you need to do is write the paragraphs that provide the content for each topic and subtopic.

A Word outline that shows just the headings for an eLearning report

- Introduction to eLearning
- Typical eLearning courses
- Why most eLearning fails
- Why our eLearning courses would succeed?
- My recommendation

A Word outline that shows the subheadings for just one heading

- Introduction to eLearning
- Typical eLearning courses
  - MOOCs
  - Commercial video courses
  - Corporate eLearning courses
  - College eLearning courses
- Why most eLearning fails
- Why our eLearning courses would succeed
- My recommendation

A checklist for reviewing a heading plan

For the headings

- Do the headings represent all the topics that the document needs to present?
- Are any of the topics unnecessary?
- Are the headings in the best presentation sequence?

For the subheadings

- Do the subheadings for each heading represent all of the subtopics that should be presented?
- Are any of the subtopics unnecessary?
- Are the subheadings for each heading in the best presentation sequence?

For heading and subheading relationships

- Is each subtopic *logically subordinate* to the topic that it's under?
- Is the *span of control* for each heading from 2 through 7?

Description

- If you're using Microsoft Word to develop a heading plan, you can use the outline feature to display just the headings for the document or the subheadings for a single heading.
- After you analyze your heading plan, you can make the required improvements.
- When you finish this step, you should feel confident that your document will be a success if you present the information that's identified by the headings and subheadings.

Figure 2-6 Step 4: Analyze and improve the heading plan

How your heading plans will affect your writing

By now, you should start to see what a difference the headings and subheadings in a document can make. But to highlight that difference, the next three figures present more examples, both good and bad.

How better heading plans will lead to better writing

Figure 2-7 starts by presenting a heading plan that has been extracted from an article in a technical magazine. This 8-page article entitled "The Best eLearning Authoring Tools" contains five headings and no subheadings."

If you analyze its headings, though, you have to wonder how "Pricing and plans" could come before you find out what the "Core features and functionality" are. And what about "Differentiating characteristics"? Shouldn't they be included in the topic on the core features? Not to mention "Some Unique Features," which you would expect to be part of the differentiating characteristics.

Well, as you might expect, an article with headings like that is probably going to be a mess throughout, and it was. In fact, when you finish it, you're hard pressed to know where to start in your search for the authoring tool that's right for you.

But what if the author had used a heading plan like the one in the *after* example? That would force the writer to introduce these tools, present both the basic and the advanced features, and only then get into the purchasing considerations. Since price alone is never the basis for buying a product, that would also get into how well the product worked in terms of both training effectiveness and developer productivity.

The second example in this figure presents the headings used for a report called "CRM Metrics that Really Work." Here again, the document uses headings but no subheadings…even though you would think that input metrics, process metrics, and output metrics would be subordinate to "How to Measure CRM Success." You also have to wonder whether the topics are in the best presentation sequence.

In fact, whenever you have more than 6 headings in a document without any subheadings, the presentation gets tedious and the relationships are hard to understand. That goes back to the span of control. What the reader wants to see is some structure that makes it easier to understand the content, not a long list that doesn't show the relationships.

In the *after* example, you can see a heading plan that would work better. It has only three headings with subheadings for the first two. If the author had used these headings, they would have forced him to present the information in a more logical and useful way…a way that would be much easier to read and understand.

Incidentally, to analyze the headings in an existing document, you usually need to page through the document and enter the headings and subheadings into a new document. That's how we created the *before* heading plans in this figure and the first two examples in the next figure. As you do that, you will often wonder how the writers decided on the headings and subheadings that they used.

Two heading plans for "The Best eLearning Authoring Tools"

Before

What are eLearning Authoring Tools
Pricing and Plans
Core Features and Functionality
Differentiating Characteristics
Some Unique Features

After

An introduction to eLearning authoring tools
- What eLearning authoring tools are
- The three types of authoring tools

Features and functionality
- Basic features
- Advanced features

Purchasing considerations
- Pricing and plans
- Training effectiveness
- Developer productivity

Two heading plans for "CRM Metrics That Really Work"

Before

What is CRM supposed to do?
CRM is based on faulty assumptions
How to Measure CRM Success
Input Metrics
Process Metrics
Output metrics

After

What is Customer Relationship Management (CRM)?
- The assumptions that prevent successful CRM
- The changed perspectives that will lead to success

How to measure the success of CRM
- Input metrics
- Process metrics
- Output metrics

How to get the most from CRM

Description

- When you extract the headings and subheadings from reports, proposals, and other business documents, you will often find that they don't identify the topics and subtopics and they don't have a structure that will help the readers see the relationships.
- That's why using a heading plan to plan what you're going to write is one of the quickest ways to improve your writing.

Figure 2-7 How better heading plans will lead to better writing

How poor heading plans will limit your effectiveness

Figure 2-8 presents three more heading plans that show how common the ineffective use of headings is. These heading plans are for an article in the *Harvard Business Review*, a chapter in a best-selling book, and a chapter in a college textbook. Over the years, I've analyzed the use of headings and subheadings in hundreds of reports and dozens of books and textbooks, and I'm sorry to say that these are all too typical of what you'll find.

The first heading plan in this figure is for an article in the *Harvard Business Review* called "The Workplace Evolution." In this plan, 8 headings are used in a 10-page article. But do the headings identify the topics and show the relationships between these topics? Do they show the flow of logic in the article? Or are the headings so vague that they could be the headings for an article on just about any business subject?

But this is just looking at it from the reader's point of view. More important is how the heading plan affects the writer. Because if your heading plan doesn't have a logical flow, doesn't identify the topics and subtopics, and doesn't show the structure between them, how can your article or chapter be logical and convincing? That's why learning how to develop an effective heading plan may just be the most important skill that you'll learn from this book.

The second heading plan in this figure shows that the same limitations often apply to the headings and subheadings used in books, even best-selling books. In fact, I extracted this example from a chapter in a best-selling book called *A World Without Work*, and I fail to see any logic or structure in the headings. Apparently, the author thinks that something is wrong with our education system, but the headings give me no clue as to what that might be. And they sure don't make it easy for me to find the facts that I might want to refer to later on.

Even with those shortcomings, though, the book is chockful of information. And if you're interested in the subject, it's worth taking the time to plow through that information. You just have to wonder how much better the book would be if the headings and subheadings forced the author to harness his facts.

The third heading plan in this figure is from a chapter on planning in a college textbook called *Introduction to Technical Writing*. This is typical of textbooks because it tries to present all aspects of the subject without presenting any of the skills that are related to it. So when you finish this chapter, you may know what the 4 types of readers, the 3 varieties of tone, and the 10 major forms of technical writing are, but you won't know how to plan what you're going to write for any of those combinations. You may also notice that this heading plan... for a chapter on planning...doesn't even mention headings and subheadings, even though headings and subheadings are critical elements in all technical writing.

In all three of these examples, the point is not so much to criticize as to show how pervasive the problem of poor heading use is. It's also to assure you that you can do better. And that starts by using the skills in this chapter to develop better heading plans.

The heading plan for an HBR article: "The Workplace Evolution"

The Value of the Modern Workplace
A Fundamentally New Approach
The Employee Engagement Crisis
Reconsidering Digital Priorities
The Barriers to Change
Tools for Teamwork
Changing Mindsets
The ROI of Workplace Transformation

The heading plan for Chapter 9: Education and its Limits from a best-selling book: *A World Without Work*

THE HUMAN CAPITAL CENTURY
WHAT WE TEACH
HOW WE TEACH
WHEN WE TEACH
THE BACKLASH AGAINST EDUCATION
THE LIMITS TO EDUCATION
THE END OF THE ROAD

The heading plan for Chapter 1: Planning from a college text book: *Introduction to Technical Writing*

- Determining the Purpose
  - A Definition of Purpose
  - Possible Purposes of Technical Writing
- Focusing on the Reader
  - Types of Readers
  - Needs of Readers
  - Number of Readers
- Limiting the Scope of the Topic
  - A Definition of Topic and Scope
  - The Scope of Professional Topics or Student Topics
- Planning the Proper Tone
  - A Definition of Tone
  - Varieties of Tone
  - Factors That Influence Tone
  - Ways to Control Tone
- Choosing a Suitable Form and Style
  - Definitions of Form and Format
  - Factors That Determine Format
  - Definition of Style
  - Ten Major Forms of Technical Writing

Description

- When you extract the heading plans from published articles, books, and textbooks, you will usually find that they have the same shortcomings that business documents have.

Figure 2-8 How poor heading plans will limit your effectiveness

The heading plans for three chapters in this book

I thought it might be interesting for you to take a step back and look at some of the heading plans that are used for the chapters in this book. So figure 2-9 presents the heading plans for the next two chapters as well as the plan for this chapter. Of course, I used the techniques in this chapter to develop these heading plans. And now, you can decide whether these heading plans will lead to effective chapters.

Of the three heading plans, the one for chapter 3 was the easiest to develop. That's because its content is the most limited. First, you need to learn the principles of paragraphing. Then, you need to learn how to provide continuity between paragraphs and sentences. I could have left it at that, but I thought you also needed to know how to write special types of paragraphs and how to plan the paragraphs that you're going to write. So I added those subtopics in a general topic named "Three related skills." That's often a good way to include odds and ends at the end of a chapter.

In contrast, the heading plan for chapter 4 was more difficult to develop. That's because there is so much material on how to write sentences. In fact, there are dozens of books that focus on just that. So the trick was to select the skills that matter the most and put them into a useful sequence and presentation structure. In the end, I broke the content down into an introduction on readability, four ways to improve readability, three ways to say what you mean, and two ways to improve your style. If that sounds simple, it wasn't. In fact, I tried several structural variations before I settled on the one in this figure.

Now, look at the heading plan for this chapter. It starts with an introduction to headings and subheadings and then presents a four-step procedure for developing heading plans. Here again, I could have left it at that, but I felt that I needed to show you some examples that prove how important the use of headings and subheadings are.

And here again, I considered several options before I settled on the final heading plan. In fact, I started with a plan that combined the skills for planning short documents with the skills for planning long documents. But in the end, I decided that would be too cumbersome, so I divided the one chapter into two: one for planning short documents (chapter 1) and one for planning long documents (this chapter).

Of course, dividing one chapter into two is something you can do when you're writing a book. But that also makes sense when you're writing some business documents. If, for example, you're planning one report that is intended for two target audiences, you may realize that it would be better to write two reports: one for each audience. That would simplify each report and help ensure its success.

The heading plan for chapter 3, *How to write paragraphs*

The principles of paragraphing
- Principle 1: Start each paragraph with the idea of the paragraph
- Principle 2: Put one and only one idea in each paragraph
- Principle 3: Fully develop the idea of each paragraph

How to provide continuity between paragraphs and sentences
- Use connecting words
- Use subject and word repetition
- Use pronouns and pointers
- Use parallel structures

Three related skills
- How to write introductory paragraphs like topic openers
- How to write paragraphs that present lists
- When and how to plan the paragraphs that you're going to write

The heading plan for chapter 4, *How to write readable sentences*

What you should know about readability measurement
- How to measure readability
- What GL scores can't measure

Four ways to improve readability
- Simplify your sentences
- Simplify your words and phrases
- Use fewer words
- Use four basic sentence structures

Three ways to say what you mean
- Be specific and prescriptive
- Use active voice
- Avoid figurative language, trite language, and analogies

Two ways to improve your style
- Write with a conversational style
- Use some stylistic devices

The heading plan for this chapter, *How to use headings and subheadings*

An introduction to the use of headings and subheadings
- How headings and subheadings should guide your readers
- Why you should use a heading plan to plan long documents

How to use a heading plan to plan long documents
- Step 1: Select and sequence the topics
- Step 2: Add and sequence the subtopics
- Step 3: Convert the topic and subtopic names to headings and subheadings
- Step 4: Analyze and improve the heading plan

How your heading plans will affect your writing
- How better heading plans will lead to better writing
- How poor heading plans will limit your effectiveness
- The heading plans for three chapters in this book

Figure 2-9 The heading plans for three chapters in this book

Perspective

How to plan the headings and subheadings for a document is one of the tested writing methods that you won't find in other books on writing. That's why the methods in this chapter will put you way ahead of most business writers. In fact, when you plan the headings and subheadings for a document as shown in this chapter, you are well on your way to writing a successful business document.

Terms

heading plan
TL;DR (Too Long; Didn't Read)
topic
subtopic
heading
subheading
first cap
title case
all caps
logical subordination
span of control
verbal

Summary

- Headings and subheadings are critical to the success of most business documents because they not only make it easier to read and understand the documents but also to refer to specific topics and subtopics later on.
- A *heading plan* can be used to plan what you're going to write. This plan consists of the *headings* and *subheadings* that identify the *topics* and *subtopics* for a document. It also shows the sequence in which the topics and subtopics will be presented.
- To create a heading plan for a document that requires both headings and subheadings, you (1) select and sequence the topics, (2) add and sequence the subtopics, (3) rewrite the topic and subtopic names as headings and subheadings, and (4) analyze and improve the heading plan.
- When you add the subtopics to a heading plan, you must be sure that each subtopic is *logically subordinate* to the topic that it's under. You should also limit the *span of control* for each topic to a maximum of 7 subtopics.
- When you rewrite topic and subtopic names as headings and subheadings, you need to be specific so the reader can tell what each topic and subtopic is going to present. You should also use consistent language structures.
- To improve your headings and subheadings, you can promise a benefit, use long headings and subheadings, and avoid *verbals* like *using* and *understanding*.
- When you use *first cap* for your headings and subheadings, instead of *title case* or *all caps*, you make a minor but professional improvement to your headings and subheadings.

3

How to write paragraphs that sell your ideas

If you analyze the writing in business documents, you'll find that one of the common flaws is poor paragraphing. That's why it's hard to follow the logic of the ideas or even to understand what the ideas are. And that's why it's so important to learn how to write effective paragraphs. This chapter will show you how.

The principles of paragraphing

This chapter starts by presenting the three principles of paragraphing. When writing is confusing or difficult to read, that's usually because the author hasn't applied one or more of these principles.

If, for example, you don't start each paragraph with the idea of the paragraph, your readers may not be able to follow the logic or flow of your writing. If you put two or three ideas in a single paragraph, your readers are going to have trouble extracting those ideas. And if you don't develop the idea of each paragraph, your readers are going to have a hard time understanding the ideas. But let's take these principles one at a time.

Principle 1: Start each paragraph with the idea of the paragraph

When you apply the first principle of paragraphing to your paragraphs, it's easier for your readers to follow the logic of your writing. In fact, your readers should be able to understand the logic or "flow" of your writing just by reading the first sentence in each paragraph. This is illustrated by the example in figure 3-1. Here, the first sentence in each paragraph is boldfaced.

If you read just those first sentences, you can see that the first paragraph is going to describe where PowerPoint slides have traditionally been used. The second paragraph is going to describe the use of slides for eLearning. The third paragraph is going to present the advantages that slides have over video. And the fourth paragraph is going to describe some of the limitations of slides. That makes it easy for the reader to follow the flow of the content.

Note, here, that the first sentence doesn't need to present the complete idea. That idea is completed and enhanced by the rest of the sentences in the paragraph. But at the least, the first sentence needs to introduce the idea. That is often done in combination with a transition from the previous paragraph.

How important is this principle? If you want to get a quick idea of how well a report or book is written, open it anywhere and start reading just the first sentences in the paragraphs. If you can follow the logic or flow of the writing without much difficulty, that's a sure sign of good writing. If you can't, the report or book is most likely going to be hard to read and understand.

Incidentally, if you read other books on writing, you'll see that some define a paragraph as a group of sentences that present a single point, some as a group of sentences that present a single thought, and some as a group of sentences that present a single subject. However, all of the books agree on the principle that each paragraph should contain only one idea, point, thought, or subject.

Since most books use the term idea for what it is that a paragraph presents, that's the term we use throughout this chapter and book. We also say that each sentence should present one and only one thought. In other words, you write paragraphs that present the ideas that consist of thoughts.

The first principle of paragraphing

- Start each paragraph with the idea of the paragraph.

Four paragraphs that start with the idea of the paragraph

For many years, PowerPoint slides have been used for business presentations, employee training, and college courses. In all of these cases, the presenter goes through the slides, one at a time, while explaining or enhancing the information on each slide.

In recent years, though, slides have become a primary medium for delivering the content in eLearning courses. In those courses, the student or trainee goes through the slides one at a time, and the slides deliver the content for the course...no instructor is required. Those slides can also deliver quizzes, animations, and more.

When you compare the use of slides with video, slides have some obvious advantages. First, you control how fast you step through the slides. Second, it's easier to skip over the information that you already know because you just press a key or click the mouse to go to the next slide. Third, it's easier to review what you can't remember because you can go quickly backward or forward through the slides to find what you're looking for.

The trouble is that slides have some serious limitations when it comes to delivering the content for subjects that are complicated. The primary limitation is that slides force you to break down the content into chunks that are so small that it's hard to see how the chunks relate to each other. But it's seeing those relationships that lead to deeper understanding and the ability to apply what you've learned. For programming subjects, this means that there's no way to present something like a lengthy program without breaking it down into pieces that don't make sense by themselves.

The test of this principle

- Your readers should be able to see the logic of your writing just by reading the first sentence in each paragraph.

Description

- For most business writing, it's best to start each paragraph with the idea of the paragraph.
- In some cases, it makes sense to start a paragraph with a transition from the previous paragraph, but even then, the sentence should hint at the idea of the paragraph.

Figure 3-1 Principle 1: Start each paragraph with the idea of the paragraph

Principle 2: Put one and only one idea in each paragraph

When you put more than one idea in a paragraph, your writing is going to be hard to read and understand. This is illustrated by the paragraph in the first example in figure 3-2. If you read it, you'll see that it's hard to understand. But why?

When you put more than one idea in a paragraph, it usually leads to two other problems. First, you fail to develop all of the ideas in the paragraph (the third principle of paragraphing). Second, you tend to mix up the sentences that relate to each of the ideas.

A third problem with paragraphs that contain more than one idea is that they are often too long. That's a problem because good readers like to pause after each paragraph to reflect on what they've learned. But when the paragraphs are more than 10 printed lines, it gets harder to digest what each paragraph is trying to say and harder to reflect on what each paragraph has said.

If that rule of thumb seems arbitrary, here's the advice of E. B. White and William Strunk, Jr. in their classic book on writing called *The Elements of Style*. They say that good paragraphing "calls for a good eye as well as a logical mind." So when you see a single paragraph that's a half-page long, let your eye tell you that you should divide it into two or more paragraphs.

This raises the question: What constitutes a paragraph idea? Or, how do you decide what a paragraph idea is? The answer is that paragraph ideas are to some extent arbitrary. As a result, more than one approach can work.

What's most important is that you have a clear view of what the idea of each paragraph is and what your flow of ideas is. Also, each paragraph idea should represent a manageable unit of information that moves the presentation forward. Then, if you write each paragraph so it obeys the principles of paragraphing, your paragraphs will be effective.

This is illustrated by the second example in this figure. Here, the paragraph in the first example has been divided into three paragraphs. Now, the reader can take a breath after each idea and more easily see how the ideas are related. That shows how important this principle is.

The second principle of paragraphing

- Put one and only one idea in each paragraph.

A paragraph that should be divided into two or more paragraphs

Video is the primary medium for delivering the content in many eLearning courses. And there's no question that video is effective whenever you can learn more by seeing how something is done than by reading about it. But video has some limitations. That starts with the fact that most people can read from 3 to 5 times faster than a video presents information. So even when video is effective, it's inefficient. Not to mention that video is available for only a small portion of the information and skills that you are likely to need. Beyond that, you can't just skip over the information in a video that you already know. It's also hard to refer to the information that you need to review.

The paragraph after it has been divided into three paragraphs

Video is the primary medium for delivering the content in many eLearning courses. And there's no question that video is effective whenever you can learn more by seeing how something is done than by reading about it. For instance, video is effective for showing how to dance, how to play tennis, and how to use the IDE (Integrated Development Environment) for a programming language.

But for the content in most courses, especially technical courses, video has some serious limitations. To start, most people can read from 3 to 5 times faster than a video presents information. So even when video is effective, it's an inefficient way to learn. Not to mention that video is available for only a small portion of the information and skills that you are likely to need.

Beyond that, you can't just skip over the information in a video that you already know. Instead, you have to watch it, or at least parts of it, to know where that information ends. Similarly, it's hard to refer back to the information that you need to review because there's no quick or easy way to find the information.

When to divide a long paragraph

- When it becomes longer than 10 printed lines.

Description

- Business writers often write paragraphs that contain more than one idea. The irony is that they often do that when they're presenting the most complicated information.
- The longer a paragraph is, the more difficult it is to read and understand. So at some point, you need to arbitrarily divide one long paragraph into two or more paragraphs

Figure 3-2 Principle 2: Put one and only one idea in each paragraph

Principle 3: Fully develop the idea of each paragraph

To "fully develop the idea of a paragraph" means to present enough supporting information to get the idea across to the readers. In other words, you support your ideas with facts, anecdotes, examples, and more. To say it another way, you support your *generalizations* with *specifics*. Not doing that is a common problem in business writing.

This problem is illustrated by the first example in figure 3-3. Here, the paragraphs all start with the idea of the paragraph and all present just one idea. However, since those ideas aren't fully developed, these paragraphs are difficult to understand. In contrast, the second example shows the difference when the ideas are fully developed.

One warning sign that this third principle hasn't been applied is a paragraph that contains less than three sentences. That's because most ideas can't be fully developed in less than three sentences. As a result, you can usually improve your paragraphs just by making sure that each one has at least three sentences.

The third principle of paragraphing

- Fully develop the idea of each paragraph.

Three paragraphs with ideas that haven't been fully developed

Perhaps the primary reason that so many eLearning courses fail is that the authoring tools force you to focus on how you're going to deliver the content. That includes the use of video, videocasts, slides, quizzes, simulations, interactions, and transitions.

But none of those eLearning features will ensure that a course will be effective. In short, there's more to teaching than technology.

A related point is that once the course is over, most students will need to enhance their skillsets. To do that, however, they usually won't be able to watch a video or play a game.

The same paragraphs after the ideas have been fully developed

Perhaps the primary reason why so many eLearning courses fail is that the authoring tools force you to focus on how you're going to deliver the content instead of the best way to deliver it. That focus is not only on how to use video, videocasts, and slides to present the content, but also on what types of quiz questions to use, when and how to use simulations, what kind of interactions and transitions to use, whether you should add gamification (and awards) to your courses…and more recently, when and how to use Virtual Reality in your courses.

The trouble is that none of those eLearning features will ensure that the course will be effective. In fact, the more the technology takes the student away from the skills that the course is supposed to teach, the less effective the course is likely to be. If, for example, the goal of the course is to teach Python programming, what the student should be doing most of the time is using a Python IDE to code, test, and debug Python programs. Anything that takes away from that… like fancy quiz questions or games…reduces the effectiveness and efficiency of the course.

A related point is that once the course is over, most of the students will need to or want to enhance their skillsets. To do that, however, they usually won't be able to watch a video or play a game. Instead, they'll have to read traditional media like books, technical manuals, or website articles. In fact, that's how most successful people continue to enhance their knowledge and skills throughout their lives. And that's one more reason why the focus on technology tends to be counterproductive…especially for college courses.

Paragraphing guideline

- To fully develop a paragraph, you usually need to write three or more sentences that provide all the information that the readers may need for understanding the idea. That may include an example, restating the idea in another way, and so on.

Figure 3-3 Principle 3: Fully develop the idea of each paragraph

How to provide continuity between paragraphs and sentences

If all your paragraphs obey the principles of paragraphing, they're going to be good paragraphs. Then, to make them even better, you should provide continuity between the paragraphs and between the sentences within the paragraphs. That will make it easier for your readers to follow the logic or "flow" of your paragraphs as well as the logic or flow of the sentences within your paragraphs. In short, your writing will be easier to read.

Use connecting words

As figure 3-4 shows, one way to provide continuity is to use *connecting words*. To illustrate, the connecting words in the example are boldfaced. In this case, the second paragraph is connected to the first paragraph by the word "However." This shows that the second paragraph will present an idea that contrasts the idea of the first paragraph.

This example also illustrates how connecting words can provide continuity between sentences. In the first paragraph, the second sentence starts with the word "And" to show that it will add to the first statement, and the third sentence starts with "For instance" to show that it will present an example that supports the first statement.

Connecting words used for continuity between paragraphs and sentences

Video is the primary medium for delivering the content in many eLearning courses. **And** there's no question that video is effective whenever you can learn more by seeing how something is done than by reading about it. **For instance**, video is effective for showing how to dance, how to play tennis, and how to use the IDE (Integrated Development Environment) for a programming language.

However, for the content in most courses, especially technical courses, video has some serious limitations. That starts with the fact that most people can read from 3 to 5 times faster than a video presents information. **So** even when video is effective, it's an inefficient way to learn. Not to mention that video is available for only a small portion of the information and skills that you are likely to need.

Besides that, you can't just skip over the information in a video that you already know. **Instead**, you have to watch it, or at least parts of it, to know where that information ends. **Similarly**, it's hard to refer back to the information that you need to review because there's no quick or easy way to find the information.

Common connecting words

Purpose	Connecting words	Purpose	Connecting words
Restate an idea	In other words	Present a summary	In short In brief In summary
Present another idea	And Besides Also In addition	Arrange ideas in sequence	First Second Third Next Then Last
Present a contrasting idea	But However On the other hand Nevertheless In contrast Instead	Arrange ideas in time	Meanwhile Later Before
Present an example	For example For instance To illustrate	Arrange ideas in space	Above Below To the left To the right Beyond
Present an idea that follows logically	So As a result Therefore		

Description

- One way to provide continuity between paragraphs and sentences is to use *connecting words*. This technique is the one that's used the most by business writers.

Figure 3-4 Use connecting words to provide continuity

Use subject and word repetition

Another way to provide continuity is to use *subject* and *word repetition*. This is illustrated by the three paragraphs in figure 3-5. Here, the repeated words are boldfaced.

Because all three paragraphs are about *video*, that word is repeated seven times in those paragraphs. Those are clear indicators of what those paragraphs are about and how the paragraphs and sentences are related. This repetition also makes it easier for the reader to follow the flow of logic from sentence to sentence and from paragraph to paragraph.

In practice, you usually don't have to deliberately use subject and word repetition. Instead, if you concentrate on the full development of each paragraph idea, that repetition will come naturally. When you're learning, though, you may want to check your writing to make sure that you're using subject and word repetition.

You should also recognize and avoid "elegant variation." The term *elegant variation* was first used in H. W. Fowler's classic reference book, *Modern English Usage*. This refers to the practice of varying a word for the sake of variation, which is contrary to the notion of using subject and word repetition. This is illustrated by the first paragraph in the second example in this figure. Here, three different words are used to refer to a report: *study*, *report*, and *document*.

In contrast, the second paragraph in this example fixes the elegant variation by repeating the word *report* three times. In this case, the result is just a bit easier to read and understand. But in some cases, elegant variation can be quite confusing, and it's a sure sign of amateur writing.

Word repetition used for continuity between paragraphs and sentences

Video is the primary medium for delivering the content in many eLearning courses. And there's no question that **video** is effective whenever you can learn more by seeing how something is done than by reading about it. For instance, **video** is effective for showing how to dance, how to play tennis, and how to use the IDE (Integrated Development Environment) for a programming language.

However, for the content in most courses, especially technical courses, **video** has some serious limitations. That starts with the fact that most people can read from 3 to 5 times faster than a **video** presents information. So even when **video** is effective, it's an inefficient way to learn. Not to mention that **video** is available for only a small portion of the information and skills that you are likely to need.

Basides that, you can't just skip over the information in a **video** that you already know. Instead, you have to watch it, or at least parts of it, to know where that information ends. Similarly, it's hard to refer back to the information that you need to review because there's no quick or easy way to find that information.

Avoid "elegant variation"

A paragraph with elegant variation

The recent **study** done by the trade association presented several conclusions that shouldn't be ignored. To start, the **report** suggests that we should stop developing Windows applications and move to web applications. Then, the **document** presents an analysis that suggests significant cost savings if we do that.

The same paragraph with subject repetition

The recent **report** done by the trade association presented several conclusions that shouldn't be ignored. To start, this **report** suggests that we should stop developing Windows applications and move to web applications. Then, this **report** presents an analysis that suggests significant cost savings if we do that.

Description

- Another way to provide continuity is to repeat the subject or key words in related paragraphs and sentences. This is called *subject repetition* or *word repetition.*
- You usually don't have to think about providing subject and word repetition because it comes naturally if your paragraphs and sentences have a logical flow.
- *Elegant variation* occurs when you deliberately vary key words, but that's a writing habit that you need to stop because it conflicts with subject and word repetition.

Figure 3-5 Use subject and word repetition to provide continuity

Use pronouns and pointers

A third way to provide continuity is to use *pronouns* and *pointers* as shown in figure 3-6. Although this technique is commonly used by professional writers, most business writers don't use this technique enough.

This is illustrated by the two examples in this figure. The first one doesn't use pronouns and pointers, but the second one does. Here, the pointer *these* and the pronouns *their*, *them*, and *they* are used. Although the difference isn't dramatic, it helps keep you moving toward a professional level of writing.

To make a reference clear, however, a pronoun or pointer needs to refer to the first noun of the same type that precedes it. In the first paragraph that uses pronouns, for example, the word *their* refers to *publishers* because *publishers* is the first plural noun that precedes it. Similarly, *they* and *their* in the second paragraph refer to *instructors*, and *them* refers to *courses*.

When you use pronouns and pointers instead of word repetition, you reduce the number of words in your sentences. That by itself makes your sentences easier to read. Even better, pronouns and pointers make the continuity between your sentences and paragraphs more obvious. That's why this is a writing skill that all business writers should add to their skillsets.

Two paragraphs that don't use pronouns and pointers

In the college environment, books and eBooks are still the primary delivery system for most courses. For instance, self-study quizzes are still printed in the end-of-chapter activities, not provided as online self-correcting quizzes. Although it's true that most large publishers also provide one video per chapter for the programming textbooks that they publish, the videos are usually of marginal value. Also, the videos are designed to enhance the textbook content, not to be the primary delivery system for the content.

In recent years, however, eLearning courses have been replacing books as the primary delivery system for many courses. And it appears that the instructors are okay with that, perhaps because the instructors don't have to take the time to develop the courses. Besides that, the courses are easier to run than traditional courses, which makes the courses especially good for adjunct instructors.

The paragraphs when pronouns and pointers are used

In the college environment, books and eBooks are still the primary delivery system for most courses. For instance, self-study quizzes are still printed in the end-of-chapter activities, not provided as online self-correcting quizzes. Although it's true that most large publishers also provide one video per chapter for **their** programming textbooks, **these** videos are usually of marginal value. Also, **these** videos are designed to enhance the textbook content, not to be the primary delivery system for the content.

In recent years, however, eLearning courses have been replacing books as the primary delivery system for many courses. And it appears that the instructors are okay with that, perhaps because **they** don't have to take the time to develop **their** own courses. Besides that, the courses are easier to run than traditional courses, which makes **them** especially good for adjunct instructors.

Common pronouns

he, she, it, they (subjective)
him, her, it, them (objective)
his, her, its, their (possessive)

Common pointers

this, that, these, those, there, here

Description

- A third way to provide continuity is to use *pronouns* and *pointers* to refer to nouns in related sentences or paragraphs. This is one of the marks of a professional writer.
- When you use pronouns and pointers, you need to make sure that the references are clear. In general, a singular pronoun or pointer refers to the first preceding noun that's singular. And a plural pronoun or pointer refers to the first preceding noun that's plural.

Figure 3-6 Use pronouns and pointers to provide continuity

Use parallel structures

One other way to provide continuity is to use *parallel structures*. These are just sentences that repeat the same language elements. This is illustrated in figure 3-7.

The first example in this figure shows how parallel structures can be used to provide continuity between sentences. In this case, the last three sentences contain this structure: "you can." If you read the same paragraph but without parallel structures, you can see how that makes the flow of sentences harder to follow. Yes, the difference between the two paragraphs is subtle, but everything helps when your goal is clarity.

In the second group of examples, you can see how parallel structures can be used to provide continuity between paragraphs. In both examples, the first sentences of the paragraphs have parallel structures.

Often, though, writers don't use parallel structures, even when the paragraphs fit that pattern. This is illustrated by the last example in this figure. In this case, the paragraphs present three ways to provide continuity, but they don't use parallel structures. Although the paragraphs still make sense, it's harder for the readers to see the continuity between them.

When you use parallel structures, you of course need to keep them as parallel as possible. That way, the readers can easily recognize the structures. The more the structures vary, the more difficult it is for the readers to see the parallels.

Parallel structures within a paragraph

A paragraph with parallel sentence structures

For most of the content of a course, video has some serious limitations. That's especially true when you compare a book or eBook to a video that has the same information. For instance, **you can read** a book 3 to 5 times faster. **You can skip over** information that you already know. And **you can refer back** to information that you want to review.

The same paragraph without parallel sentence structures

For most of the content of a course, video has some serious limitations. That's especially true when you compare a book or eBook to a video. For instance, it takes 3 to 5 times longer to watch a video than it takes to read a book that has the same information. Book information can easily be skipped if you already know it. And there's no good way to refer back to video information if you want to review it.

Parallel structures that start a series of paragraphs

Example 1

The first way to provide continuity is to …

The second way to provide continuity is to …

The third way to provide continuity is to …

Example 2

Video casts are used to …

Slideshows are used to …

Simulations are used to …

Paragraph beginnings that would be better if they were parallel

One way to provide continuity is to use …

Subject and word repetition is a second way …

An elegant way to provide continuity is to …

Description

- When you use *parallel structures* to start a series of paragraphs, you help the readers see the relationships between the paragraphs.
- When you use parallel structures within a paragraph, you help the readers see the relationships between the sentences.
- When you use parallel structures, be sure to keep them parallel. The more they vary, the harder it is for your readers to see the parallels.

Figure 3-7 Use parallel structures to provide continuity

Three related skills

In general, the principles of paragraphing apply to all the paragraphs that you write. But the subtopics that follow present some ideas for writing special types of paragraphs.

How to write introductory paragraphs like topic openers

Figure 3-8 presents some ideas for writing introductory paragraphs. To start, you need to know that the paragraph or paragraphs that introduce a series of subtopics is called a *topic opener*. Often, this type of opener consists of a single paragraph that requires just a sentence or two. This is illustrated by the topic opener in the first example in this figure. This is one of the exceptions to the principle that each paragraph should be fully developed, usually by three sentences or more.

Similarly, a paragraph that introduces a series of paragraphs doesn't have to be fully developed. This is illustrated by the second example in this figure. Here, a two-sentence paragraph introduces the paragraphs that present the limitations of video instruction. This is followed by one paragraph for each limitation.

Then, the third example shows that it's also okay to combine an introductory paragraph with the first paragraph in a series. In that case, the introductory paragraph can be seen as fully developed. But either way is acceptable for business writing.

A topic opener in an eLearning report

A quick tour of some typical eLearning courses

Who offers eLearning courses? Dozens of websites and corporations. Here's a quick tour of some of the major providers. — **Topic opener**

MOOCs

A MOOC (Massive Open Online Course) is an online course that's offered free from a website. For most MOOCs, the primary method for delivering content is video lecture. However, many MOOCs also provide quizzes as well as user forums that support interactions among students and

A paragraph that introduces other paragraphs

For most of the content of a programming course, video has serious limitations. Here are the most important ones.

First, most people can read from 3 to 5 times faster than a video presents information. So even if a video is effective, it's an inefficient way to learn. That's why most video courses let you speed the video up by 1.5, 2, or even 3 times the normal playing speed. But the faster the video plays, the harder it is to grasp the content.

Second, when you play a video, it's hard to skip over information that you already know. That slows you down even more because you have to go through information that you already know to get to the new information. Compare that to skimming through the pages in a book.

Third, …

An introductory paragraph with the first idea embedded within it

For most of the content of a programming course, video has serious limitations. First, most people can read from 3 to 5 times faster than a video presents information. So even if a video is effective, it's an inefficient way to learn. That's why most video courses let you speed the video up by 1.5, 2, or even 3 times the normal playing speed. But the faster the video plays, the harder it is to grasp the content.

Second, …

Third, …

Description

- A *topic opener* introduces the subtopics that follow. For many topics, the opener will require just one paragraph, and that paragraph doesn't have to be fully developed. As a result, it can consist of just a sentence or two.
- Similarly, a paragraph that introduces a series of other paragraphs doesn't have to be fully developed so it can consist of just a sentence or two. In some cases, though, it makes sense to embed the first item in the series within the introductory paragraph.

Figure 3-8 How to write introductory paragraphs like topic openers

How to write paragraphs that present lists

Figure 3-9 shows several ways to write paragraphs that present lists. The first two examples show two ways to handle *numbered lists*. In the first example, the items are embedded in the paragraph. In the second example, they are set off from the paragraph. That way your readers can quickly refer to the numbered items. In contrast, the third example is a *bulleted list* that is set off from the paragraph.

If the items in a set-off list are complete sentences, use a period at the end of each item as in the bulleted list. If they aren't complete sentences, don't use periods as in the second example. If the items in a list are mixed (some are complete sentences, some aren't), use a period at the end of each item. But try to avoid mixed items.

When you present lists, you should try to use parallel structures for the items in the list. That makes the items easier to read. Also, when you set off a list from the text of a paragraph, bulleted lists usually work as well as numbered lists. The only time you need to number the items is when there is some reason for numbering them as in the steps of a procedure or the items in a prioritized list.

When you use a bulleted list, you imply that the order of items isn't sequential. Nevertheless, the sequence of items in an unnumbered list should be as logical as possible. If, for example, you're listing benefits, you should list the items in the sequence that you think represents their decreasing order of importance.

Whenever you use lists, you should try to limit the number of items in them. That's because any list over 7 items starts to be counterproductive. This is illustrated by the last list in this figure, which contains 11 items. This list is excerpted from a book on writing, and the list actually contains 19 items.

The first problem with a list like this is that the mind starts to boggle when you reach 7 items. Some other common problems are (1) overlap and duplication of items, (2) vague or imprecise items, (3) items that don't belong in the list at all, (4) items that don't have parallel structures, and (5) items that aren't in a logical sequence.

So here's some advice on using lists. First, watch out for lists that consist of more than six or seven items. Second, make sure all of the items in the list pertain to the purpose of the list. Third, be sure to combine duplicates and eliminate overlap in the list. Fourth, use parallel structures whenever you can. Fifth, put the items in the list in some logical sequence because any order is better than no order at all.

A numbered list that's embedded in a paragraph

If you take a few eLearning courses, you'll find that they consist of just a few components. The primary components are (1) videos and videocasts, (2) slides, (3) simulations, and (4) quizzes.

A numbered list that is set off from the text of a paragraph

If you take a few eLearning courses, you'll find that they consist of just a few components. The primary components are:

1. Videos and videocasts
2. Slides
3. Simulations
4. Quizzes

A bulleted list that is set off from the text of a paragraph

For most eLearning courses, video has some serious limitations. Here are some of them:

- Most people can read from 3 to 5 times faster than the speed at which a video presents information.
- It's hard to skip over information that you already know.
- It's hard to refer back to the information that you need to review.

A bulleted list with more than 7 items...something that should be avoided

Because writing an introduction to a report can be difficult, here are 11 ways to start an introduction:

- Scope
- Point of view
- Purpose
- Problem
- Background
- Quotation
- Question
- Comparison
- Illustration
- Action
- Humor

Guidelines for presenting lists

- Try to use parallel structures for the list items.
- Unless the items in a list should be numbered, use a bulleted list.
- Try to limit the number of items in your lists to 6 or 7.

Figure 3-9 How to write paragraphs that present lists

When and how to plan the paragraphs that you're going to write

After you use a heading plan to plan the topics and subtopics that you're going to write, you write the paragraphs that present the content for those topics and subtopics. Most of the time, you don't need to plan those paragraphs because the ideas for the paragraphs will come to you as you write. So you just write one paragraph after another until you've presented all the ideas for the topic or subtopic.

But when a topic or subtopic is complicated, the sequence of ideas that you need to write may not be apparent. Then, it makes sense to plan the sequence of paragraphs before you write them. You may also find that planning is useful when you're learning how to write effective paragraphs.

To plan the paragraphs, you can just list the paragraph ideas as shown in figure 3-10. Here, the writer has written one line for each of the planned paragraphs using whatever notation she prefers. In this case, the first paragraph will be an introduction to the use of video. The next three paragraphs will compare reading to video, show that it's hard to skip over video content or refer back to it, and present a personal example. The last paragraph will draw some kind of conclusion.

As you write the paragraphs that you've planned, you may decide that you need to modify the plan. That's illustrated by the five paragraphs in this figure, which don't quite adhere to the plan. Here, the fourth paragraph starts the personal example, but the writer decided that she needed to continue that example in a fifth paragraph. She also realized that the fifth paragraph could be used to draw the conclusion.

Whether you plan the paragraphs or just start writing them, you of course need to decide what ideas you need to present. But you also need to decide what the best sequence is for presenting those ideas. For simple topics or subtopics, those decisions may be easy. But for complicated ones, those decisions may not only require planning but also some adjustments to your headings as you write the paragraphs.

A plan for the paragraphs in a topic called "The limitations of video"

Introduction to video

How video compares with reading

It's hard to skip over video content or refer back to it

Personal example

Conclusion

The paragraphs for the plan with a minor adjustment to the plan

Video is the primary medium for delivering the content in many eLearning courses. And there's no question that video is effective whenever you can learn more by seeing how something is done than by reading about it. For instance, video is effective for showing how to dance, how to play tennis, and how to use the IDE (Integrated Development Environment) for a programming language.

But for the content in most courses, especially technical courses, video has some serious limitations. That starts with the fact that most people can read from 3 to 5 times faster than a video presents information. So even when video is effective, it's an inefficient way to learn. Not to mention that video is available for only a small portion of the information and skills that you are likely to need.

Beyond that, you can't just skip over the information in a video that you already know. Instead, you have to watch it, or at least parts of it, to know where that information ends. Similarly, it's hard to refer back to the information that you need to review because there's no quick or easy way to find the information.

Because those video limitations seem so obvious, I decided to see for myself whether they were true. So, I started a LinkedIn Learning course on how to use some of the advanced features of PowerPoint. But after 30 minutes of deadly slow progress toward learning the first feature, I decided I had enough, and I added "boring" to my list of video shortcomings.

I then ordered a book on PowerPoint that in 30 minutes showed me how to use all the advanced features that I was interested in. In short, the book made it easy to skip the features that I already knew, find the features that I wanted to learn, and review the details of features as I tried to use them. For me, that not only demonstrated the limitations of video instruction, but also the power of a book or eBook.

Description

- Most of the time, you shouldn't need to plan the paragraphs that you're going to write. The paragraph ideas should come to you just as they would if you were explaining something to someone in person.
- But if the subject is complicated and the paragraph ideas don't just come to you, it may help to plan the paragraphs that you're going to write. To do that, you can use your own notation to list the ideas that you think you need to include.

Figure 3-10 When and how to plan the paragraphs that you're going to write

Perspective

In the real world of business writing, poor paragraphing is common. One problem is that the paragraphs don't start with the ideas of the paragraphs so it's hard for the readers to follow the flow of ideas. Another problem is that writers put more than one idea in their paragraphs so it's hard to separate and extract the ideas. The third problem is that writers don't fully develop their ideas so it's hard to understand them.

To solve these problems, you just need to apply the principles of paragraphing. Then, if you add continuity to your paragraphs by using the four techniques in this chapter, your paragraphs are going to be even better. At that point, you will have all the skills that you need for writing the paragraphs that will sell your ideas.

Terms

connecting word
subject repetition
word repetition
elegant variation
pronoun
pointer
parallel structures
topic opener
numbered list
bulleted list

Summary

- The three principles of paragraphing are (1) start each paragraph with the paragraph idea, (2) put just one idea in each paragraph, and (3) fully develop the idea in each paragraph.
- To provide continuity between paragraphs and sentences you can use (1) *connecting words*, (2) *subject* and *word repetition*, (3) *pronouns* and *pointers*, and (4) *parallel structures*.
- Some introductory paragraphs, like *topic openers* and paragraphs that introduce other paragraphs, don't need to be fully developed.
- When you write paragraphs that present numbered or bulleted lists, you should limit the number of items to 6 or 7. You should also try to use parallel structures for the list items and put them in a logical sequence.
- For most writing, you shouldn't have to plan the paragraphs that you're going to write. But when you're learning or when a topic or subtopic is complicated, planning the paragraphs may help. To do that, you just list the paragraph ideas with whatever notation you prefer.

4

How to write sentences that are easy to read and understand

If you adhere to the principles of paragraphing, as shown in chapter 3, you're going to write paragraphs that deliver your ideas. But your paragraphs will do that even better if you write sentences that are easy to read and understand. So that's what this chapter will teach you how to do.

To start, this chapter will show you how to write sentences that are easy to read. But that's not enough. So next, it will show you how to make sure that your sentences say what you mean. Last, this chapter will show you how to develop a writing style that your readers will appreciate. In fact, you can think of this chapter as a "crash course" on writing sentences because it presents the best practices taken from dozens of books on writing.

What you should know about readability measurement

Rudolf Flesch is the father of *readability*. In his 1951 book, *How to Test Readability*, he presented his formula for the Flesch Reading Ease Score. That score is a number from 0 to 100 that indicates how difficult (0) or easy (100) it is to read a document.

Later, Mr. Flesch teamed up with J. Peter Kincaid, a scientist and educator, to develop the Flesch-Kincaid Readability Test for the U. S. Navy. That test returns a *grade level (GL) score* that indicates the grade level in United States schools that the selection is appropriate for. This test became the U. S. standard for measuring readability in 1980. In this chapter, we use GL scores to measure readability because that score is easy to understand.

How to measure readability

Figure 4-1 starts by presenting the formula for calculating the GL score for a document. If you study this formula, you can see that the critical factors are the average sentence length and the average number of syllables in a word. So if you keep your sentences short and your words simple, you're going to get a low GL score. And that of course is what you want.

If you use Microsoft Word, you can get the GL score for a document by running its Spelling and Grammar Checker, as shown in chapter 10. If you don't use Word, you can search the Internet for programs that provide readability ratings. There, you'll see that there are several different ways to rate the readability of a document.

To illustrate GL scores, the first example in this figure is from one of Rudolf Flesch's own books. Because it has a GL score of 7.1, anyone who has finished 7th grade should be able to read it with relative ease. But keep in mind that low GL scores also work for college graduates because that makes it easier for them to get the information that they're looking for. In other words, your GL score doesn't have to be high just because you're writing for an educated audience.

In contrast, the second example has a GL score of 13.5, and the third example has a GL score of 19.4, even though they're from the same white paper on eLearning. If you read these paragraphs, you'll start to appreciate the differences in reading scores and readability. Neither one of the paragraphs is easy to read, but the second one is easier to read than the third one.

Incidentally, when I got the GL scores for excerpts from some of the best-selling business books, I found scores that ranged from grade 11 to 15. And when I rated ten business documents, I found scores that ranged from 11.2 to 26.8! Although a GL score of 12 may be acceptable for some business documents, you surely have to question anything that's much above that.

In case you're interested, the grade level score for each of the chapters in this book is 10 or below. This chapter has a GL score of 8.3. And the five paragraphs above this one have a GL score of 9.4.

The formula for calculating the grade level (GL) score

```
Flesch-Kincaid grade level score = (.39 x ASL) + (11.8 x ASW) - 15.59
```

where

```
ASL = Average Sentence Length (number-of-words / number-of-sentences)
ASW = Average Number of Syllables Per Word
```

Example 1: From *Why Johnny Can't Read* by Rudolf Flesch (GL=7.1)

It used to be the typical American ideal—and practice—to give children a better education than their parents had had. Fathers who never got beyond eighth grade sent their children to high school; high-school graduates proudly watched their sons getting college degrees. But things have changed in the last ten, twenty years. For the first time in history American parents see their children getting less education than they got themselves. Their sons and daughters come home from school and they can't read the newspaper; they can't spell simple words like February or Wednesday; they don't know the difference between Austria and Australia.

Example 2: From a white paper on eLearning (GL=13.5)

One of the missteps eLearning designers make is using technology that overpowers the design. Examples of this can be interactions that do not contribute to the learner meeting the instructional goals, or even technology that makes the learner experience more difficult. What's important is sound design that clarifies to learners what the expectations and goals are and helps them achieve those goals, regardless of the modality.

Example 3: From the same white paper on eLearning (GL=19.4)

With the shift from face-to-face learning methods to more asynchronous, eLearning methods, more designers are using technology at the various stages of systematic design than ever before. While technology can be a blessing for growing geographically diverse learner bases, it is imperative that the technology be used in a way that complements the learning and aids training and learning teams in providing evidence of its effectiveness.

Description

- Rudolf Flesch is the father of *readability* measurement. In his 1951 book, *How to Test Readability*, he presented the formula for the Flesch Reading Ease score. That test returns a number from 1 to 100 that indicates how readable a document is with 100 the easiest.
- J. Peter Kincaid is a scientist and educator who helped develop the Flesch-Kincaid Readability Test for the U. S. Navy. That test returns a *grade level (GL) score* that indicates the grade level in U. S. schools that is appropriate for reading the document.
- Today, you can use Word's Spelling and Grammar Checker to get the GL score for a document, as shown in chapter 10.

Figure 4-1 How to measure readability

What GL scores can't measure

Whenever you measure the readability of a selection, you should realize the limitations of the score. One problem is that it is the score for the entire selection. But individual paragraphs may have scores that are three or more grades above or below that score. Worse, the introductory paragraphs, the concluding paragraphs, and the paragraphs that present the most difficult content are likely to have the highest GL scores.

This is illustrated by the excerpt in figure 4-2. Here, the GL score for both paragraphs is 11.3, but the two paragraphs have individual scores of 13.6 and 9.2. That means that the first paragraph is okay for a second-year college student, and the second paragraph is okay for a tenth-grade high school student. This variation in scores within the same document is typical of business writing, and the variation is often worse than that.

For instance, the range of scores for the paragraphs in two white papers that I recently scored were from 7.1 to 16.3 with an overall score of 11.1, and from 8.1 to 17.3 with an overall score of 13.3. And, yes, the paragraphs that presented the most difficult content had the highest GL scores.

The second problem with readability scores is that a low score doesn't mean that your readers will understand what you've written. Way back in 1955, for example, one study showed only an 8 percent increase in reader comprehension when a manual was rewritten from a GL score of 16 to a GL score of 12, and no further improvement when it was rewritten to a GL score of 8. Since then, many other studies have shown that readability scores don't always correlate with reader comprehension.

The third problem with readability scores is that they don't tell you anything about the content. Is the content interesting? Is it useful? Is it convincing? Does it present the information that the readers need to know? Does it make sense? That's the real problem with readability scores. You can get low GL scores and still write documents that don't achieve their goals.

This is illustrated by the example in this figure. Here, the overall grade level score is 11.3, which should be okay for high school graduates, but it's hard to make any sense of the content. "Planning identifies the output required, while budgeting identifies the input required." What does that mean? "The relationship between planning and budgeting is dynamic." What does that mean? You might think I made this example up, but this is an excerpt from a white paper.

But if readability scores don't ensure comprehension or clarity, what does? The answer is: headings and subheadings, paragraphing, and all the other skills that you'll learn in this book. In fact, one of the reasons why the excerpt in this figure doesn't make sense is that the paragraphs don't adhere to the principles of paragraphing.

On the other hand, that doesn't mean you should disregard readability scores. In fact, if your scores are above 12.0, you really do need to improve the readability of your writing. And your writing will be even better if your scores are 10.5 or below.

A low GL score doesn't assure comprehension (GL=11.3)

Many managers have suggested that zero-base budgeting be renamed zero-base planning or maybe zero-base planning and budgeting. That's because the process requires effective planning, and it shows up any lack of planning. In fact, planning and budgeting process can be contrasted as follows. Planning identifies the output required, while budgeting identifies the input required. Planning establishes programs, sets goals and objectives, and makes basic policy decisions, while budgeting analyzes the many functions or activities that the organization must perform to implement each program. Budgeting also analyzes the alternatives within each activity to achieve the end product desired. And budgeting also identifies the trade-offs between partial or complete achievement of the established goals and the costs.

The relationship between planning and budgeting is dynamic. That's because the resources required to achieve the desired goals are limited. Therefore, we must determine whether achieving the last 10% of each goal requires 25% of the cost. Or whether we can achieve each goal. We must also decide whether we must eliminate and/or reduce some goals. If we fixed our goals, the zero-base budgeting process would be a suboptimization tool. And that tool would tell us how best to achieve the given results. However, the requirement to modify goals based on a cost/benefit analysis changes that. It makes the zero-base budgeting process both a suboptimization and total-optimization tool.

The GL scores of the paragraphs above

- 13.6 and 9.2 with an overall score of 11.3

A reasonable GL score for most business documents

- 10.5 or lower

What GL scores can't measure

- Whether the readers will be able to comprehend the information
- Whether the information makes sense

Description

- In practice, the GL scores for the individual paragraphs in a document may be two or three grade levels above or below the score for the entire document. And the paragraphs with the highest scores are likely to be the ones that present the most difficult content.
- Remember that the readability score is just an indicator of how easy it is to read a document. But the success of the document also depends on its use of headings and subheadings, its paragraphing, its use of visual aids, and all the other skills in this chapter and book.
- Nevertheless, you shouldn't disregard readability measurement. If a document has a GL score of much more than 10.5, you should probably improve its readability.

Figure 4-2 What GL scores can't measure

Four ways to improve readability

To lower your grade level scores, all you need to do is shorten your sentences and simplify your words. But there's a right way and wrong way to do that. So what follows are four ways to not only improve your GL scores, but to actually make your sentences easier to read and understand.

Simplify your sentences

To simplify your sentences, you first need to think about dividing any sentences that are longer than 20 words or about two lines of text into two or more sentences. This is illustrated by the first group of examples in figure 4-3.

In example 1, a sentence that is 40 words and more than three lines long is divided into three sentences. That drops its GL score from 21.1 to 10.7. In example 2, a sentence that is 38 words and more than three lines long is divided into two sentences. That drops its GL score from 21.8 to 13.2. In both cases, the improved GL scores don't seem to clarify the content, but at least that's a first step toward clarity.

Another way to think about simplifying sentences is to identify structures that aren't simple and then fix them. That's illustrated by the next example in this figure. Here, the original sentence includes a *non-restrictive clause*. That's a clause that isn't essential to the meaning of the sentence, and that type of clause should be set off from the rest of the sentence by commas.

The problem with a non-restrictive clause or phrase is first that it interrupts the main thought of the sentence, and second that it adds a second thought to the sentence. To fix this problem, though, you just need to take the non-restrictive clause out of the sentence. Then, if you think it presents useful information, you can convert it to its own sentence. In some cases, though, you may decide that the information isn't needed.

In the improved version of this example, the GL score drops from 14.4 to 8.8, even though all of the information is kept. And this time, the change does improve the clarity of the content.

Another sentence structure that you should avoid is a sentence that has an *introductory participial phrase*. This is illustrated by the last example in this figure. Since that type of phrase doesn't have a subject, it's hard to tell who is doing what. Worse, participial phrases often present thoughts that aren't essential to the meaning of the sentence.

In the improvement, the participial phrase is converted to a sentence of its own, and the one sentence is divided into two. That drops the GL score from 15.4 to 8.4. And here again, this change does improve the clarity of the content.

Divide long sentences into two or more sentences

Example 1: Original (GL=21.1)

While technology can be a blessing for growing geographically diverse learner bases, it is imperative that the technology be used in a way that complements the learning and aids training and learning teams in providing evidence of its effectiveness.

Improvement (GL=10.7)

Technology can be a blessing for growing geographically diverse learner bases. But it is imperative that the technology be used in a way that complements the learning and aids training. It must also provide evidence of its effectiveness.

Example 2: Original (GL=21.8)

The speed and scale at which technological breakthroughs are emerging have no historical precedent and have created an imperative for businesses across industries to respond rapidly with their own digital transformations to drive growth and create competitive advantage.

Improvement (GL=13.2)

Today, technological breakthroughs are emerging at a speed and scale that has no historical precedent. That in turn means that businesses need to rapidly respond with their own technological transformations if they want to continue to be competitive.

Avoid non-restrictive clauses and phrases

Original (GL=14.4)

Our new agency program, *which has the support of all managers and which will be the basis for all future sales activities*, represents a much-needed change in our policies.

Improvement (GL=8.8)

Our new agency program represents a much-needed change in our policies. It will be the basis for all future sales activities, and all of our managers support it.

Avoid introductory participial phrases

Original (GL=15.4)

Perceiving the switch to remote work from home as a brief emergency, the corporate leaders focused on accomplishing the immediate tasks of their organizations.

Improvement (GL=8.4)

At first, the corporate leaders thought the switch to remote work would be brief. As a result, they focused on accomplishing the immediate tasks of their organizations.

Description

- The easiest way to simplify your sentences is to divide long sentences into shorter ones.
- To further simplify your sentences, you should avoid using *non-restrictive clauses* and *phrases* as well as *introductory participial phrases.*

Figure 4-3 Simplify your sentences

Simplify your words and phrases

Figure 4-4 presents five ways to simplify your words and phrases. All of them are easy to do, and all of them will not only improve your GL scores, but also make your sentences easier to read.

First, use the simple word or expression whenever possible, as shown in the partial list of words and expressions at the start of this figure. Even in technical books, you can say "do" instead of "perform" and "run" instead of "execute." And please say "use" instead of "utilize," "finish" instead of "finalize," and "end" instead of "terminate."

Second, along the same lines, avoid using words with prefixes and suffixes. In particular, you should look for prefixes like *pre*, *re*, and *de*, and suffixes like *able*, *ation*, *ality*, *ability*, *ness*, and *tion*. So "decide" is better than "make a decision," "indicate" is better than "give an indication," and "easier to read" is better than "more readable." In this chapter, though, I've used *readable* and *readability* because they're commonly used in the context of writing.

Third, avoid using *noun clumps*. These are a series of nouns with the first nouns in the series used as adjectives, as in "software development industry." Here, all three words are nouns, but the first two are used as adjectives.

Even though noun clumps are commonly used in business writing, it's better to rewrite all clumps of three or more nouns, and even some two-word clumps. Along the same lines, you should also avoid *big-word clumps* like "highly productive software development industry" because they always make reading more difficult.

Fourth, avoid using technical jargon and acronyms. Of course, you can use the jargon and acronyms that your audience is familiar with. But if there's any doubt, you need to explain the jargon and identify the acronyms, even if that increases the grade level score.

Last, avoid the use of constructions like *and/or*, *he/she*, or worse. Sometimes, when you do that, the result will require more words. But even with that, it will be easier to read and understand.

Use the simple word or expression

Avoid	Use	Avoid	Use
in the event of	if	in order to	to
at this point in time	now	due to the fact that	because
prior to	before	following	after
utilize	use	pertaining to	about
call your attention to	point out	be of assistance	help

Avoid words with prefixes and suffixes

Original (GL=13.9)

A study done by our purchasing department gives an *indication* that our expenses for office supplies are *reducible* by as much as 15 percent.

Improvement (GL=11.3)

A study done by our purchasing department shows that we can reduce our expenses for office supplies by as much as 15 percent.

Avoid noun clumps and big-word clumps

Original (GL=18.6)

The focus of the business administration core content curriculum is competency development in management and accounting.

Improvement (GL=13.7)

The goal of the major in business administration is to develop the critical skills for management and accounting.

Avoid technical jargon

Original (GL=11.5)

The purpose of *SCORM* is *interoperability*. It *seamlessly integrates* the content of an eLearning course with an *LMS*, and it's *developer-friendly.*

Improvement (GL=13.3)

SCORM is the standard that lets a Learning Management System deliver an eLearning course without any extra work by the course developer.

Avoid constructions such as *and/*or, *he/she*, and *item(s)*

Original (GL=8.5)

When *he/she* sends the *requested/required* information, the *item(s)* will be processed.

Improvement (GL=5.8)

When you send the required item or items, your request will be processed.

Description

- If you spot any of the usages above in your writing, it's relatively easy to fix them.

Figure 4-4 Simplify your words and phrases

Use fewer words

If you use fewer words, you will of course improve the readability score of your writing. Figure 4-5 presents five ways to do that.

First, keep your verb forms simple. When you find yourself wondering what the correct verb form should be (like "should have had"), rewrite and simplify. For most technical writing, you'll find that the present tense works most of the time so you don't need to use words like "could," "should," "would," or even "will."

Second, as you learned in chapter 3, replace unnecessary word repetitions with *pronouns* and *pointers*. You can identify the need for this when you see the same words repeated as in the second example in this figure. By replacing the word repetitions with pronouns or pointers, you not only shorten the sentences but also provide better continuity between them. In fact, the use of pronouns and pointers is one of the marks of a professional writer, and most business writers don't use them enough.

Third, delete unnecessary adjectives and adverbs. When you write the first draft of a document, you typically include adjectives and adverbs that don't add to the meaning of the sentence. Then, when you edit your first draft, you can strike them out. This is especially true for adverbs.

Fourth, delete other unnecessary words, which editors call *deadwood*. Here again, when you write the first draft, you typically include all of the qualifying phrases that you think are necessary to make the sentences precise. But when you edit, you should realize that many of these qualifiers aren't needed.

This is illustrated by the fourth example. Here, the original was taken from an actual business document, and the rewrite lowers the grade level score by almost 50 percent. That's quite a difference, but one that's typical of all too much writing. You would think the writers were getting paid by the word.

When you write the first draft of a document, though, it's often hard to determine whether an adjective, adverb, or some of the deadwood is necessary. That's why it makes sense to include the adjectives, adverbs, and deadwood without giving them too much thought. Then later, when you edit the document, you can read the sentences with and without the extra words and decide whether they're needed.

Use simple verb forms

Watch out for "could," "would," and other complex verb forms (GL=4.4)

If you *would use* templates, you *wouldn't have* to start each document from scratch.

The present tense works best in most cases (GL=4.1)

If you use templates, you don't have to start each document from scratch.

Use pronouns and pointers to avoid word repetition

Original (GL=7.2)

So far, the *instructors* who use the eLearning *course* on HTML are satisfied with the *course*. I think that's because the *instructors* don't have to take the time to develop the *course* themselves.

Improvement (GL=6.3)

So far, the *instructors* who use the eLearning *course* on HTML are satisfied with *it*. I think that's because *they* don't have to take the time to develop *it* themselves.

Delete unnecessary adjectives and adverbs

Example 1: Original (GL=11.9)

To correct this problem, *simply* delete the *several* unnecessary words.

Improvement (GL=9.6)

To correct this problem, delete the unnecessary words.

Example 2: Original (GL=12.7)

Today, technological breakthroughs are emerging at a speed and scale that has no *historical* precedent. That means that businesses need to *rapidly* respond with their own *technological* breakthroughs if they want to continue to be competitive.

Improvement (GL=9.7)

Today, technological breakthroughs are emerging at a speed and scale that has no precedent. That means that businesses need to respond with their own breakthroughs if they want to continue to be competitive.

Delete "deadwood"

Original (GL=14.4)

As this report shows, our salesmen have dealt with *the matter of* competing products in many *creative and original* ways *in many different environments*.

Improvement (GL=7.5)

As this report shows, our salesmen have dealt with competing products in many ways.

Description

- *Deadwood* refers to words and phrases that don't add to the meaning of a sentence.
- To test whether adjectives, adverbs, or deadwood add to the meaning of a sentence, read the sentence without them.

Figure 4-5 Use fewer words

Use four basic sentence structures

Have you ever thought about the sentence structures that you use? Well, some work better than others...especially for business writing. That's why we recommend that you use the four sentence structures in figure 4-6 for most of the sentences you write. The reason these structures work so well is that they use subjects and verbs to deliver their meaning.

The first structure is the *simple sentence*. This is a sentence that is based on a subject and a verb. Note, however, that simple sentences aren't always short, or even simple, because they can start with introductory words and phrases. And the verbs in those sentences can be followed by objects, phrases, clauses, and more.

The second structure is the *compound sentence*. It combines two simple sentences by connecting them with *and*, *or*, or *but*. This structure is useful when you want to connect two closely related thoughts.

The third structure is a sentence with an introductory *adverbial clause*. This is a powerful sentence structure because it shows relationships. Without getting into too much grammar, a *clause* contains a subject and a verb, and an introductory adverbial clause is a *subordinate clause* that modifies the *main clause*.

Adverbial clauses start with words called *conjunctions*. For instance, conjunctions like *because* and *since* introduce clauses that tell why. Conjunctions like *after*, *before*, *until*, *while*, and *when* introduce clauses that tell when. And conjunctions like *although*, *if*, and *unless* introduce clauses that present conditions.

The fourth structure is a sentence with a concluding adverbial clause. Here again, the adverbial clause modifies the main clause, and it shows the relationship between the two clauses. But this time, the adverbial clause is after the main clause.

You can start any of these sentence structures with introductory words or phrases followed by a comma. In fact, that's a good way to counteract the deadening effect of several sentences in a row that start with the subject. Those introductory words and phrases can also be used to provide continuity between the sentences and paragraphs, as you learned in the last chapter.

The last example in this figure is an excerpt taken from earlier in this chapter. It shows that you can write any document with just these four structures. In fact, almost every sentence in this book uses one of these structures.

In this example, the first paragraph uses two simple sentences followed by one compound sentence. Then, the second paragraph starts with a sentence that has an introductory adverbial clause. And that's followed by a sentence that has a concluding adverbial clause (and also an introductory phrase).

As you use these structures, you will of course want to limit their length and complexity. And to do that, you should divide any sentence of more than 20 words or two lines of text. That will keep the GL score down, and the sentence structures will keep the clarity up. In this example, the GL score is just 8.7.

Use four basic sentence structures

Simple sentences

Each *simple sentence* consists of one independent clause. An *independent clause* contains a subject and verb. The verb can be followed by an object, phrases, clauses, and more. As a result, simple sentences aren't necessarily short.

Compound sentences

Compound sentences consist of two independent clauses, and these sentences can be used to connect two closely related thoughts.

Sentences with introductory adverbial clauses

Because sentences with introductory adverbial clauses show relationships, they are often easier to read and understand than simple or compound sentences.

Although sentences with introductory adverbial clauses tend to be longer, they do an excellent job of showing relationships.

Sentences with concluding adverbial clauses

Sentences with concluding adverbial clauses often improve readability *because they do a better job of showing relationships.*

Sentences with concluding adverbial clauses do an excellent job of showing relationships, *even though they tend to be longer.*

But avoid the deadening effect of a series of sentences that start with the subjects

- To do that, start some sentences with introductory words, phrases, or adverbial clauses.

An excerpt from this chapter that uses just these structures (GL=8.7)

Third, avoid using *noun clumps*. These are a series of nouns with the first nouns in the series used as adjectives, as in "software development industry." Here, all three words are nouns, but the first two are used as adjectives.

Even though noun clumps are commonly used in business writing, it's better to re-write all clumps of three or more nouns, and even some two-word clumps. Along the same lines, you should also avoid *big-word clumps* like "highly productive software development industry" because they always make reading more difficult.

Description

- A *simple sentence* contains a subject and a verb, but it isn't necessarily short.
- A *compound sentence* consists of two independent clauses connected by *and*, *or*, or *but.*
- Sentences that have either introductory or concluding *adverbial clauses* work well for showing relationships. And most business writers don't use them enough.
- Since a clause contains a subject and verb, all four of these sentence structures are driven by subjects and verbs. That's why they work so well.

Figure 4-6 Use four basic sentence structures

Three ways to say what you mean

As you learned earlier, low readability scores don't ensure that your readers will understand what you've written. In fact, they don't even ensure that your writing will make sense. That's why you will now learn three ways to make sure that you're saying what you mean.

Be specific and prescriptive

One of the common weaknesses of business writing is that the writers aren't *specific*. This is illustrated by the first part of the first example in figure 4-7. Here, the writer is saying that you need to write up to ten pages of documentation to get a purchase approved, but that's anything but specific.

In contrast, the rewrite is specific. It tells you how many pages of documentation you need for two levels of purchases. Of course, you would like to be more specific than that. But you get the idea. Generalities don't give the readers the information that they need.

If you're trying to show someone how to do something, it also helps to be *prescriptive*, not *descriptive*. This is illustrated by the second example in this figure. Here, the before example *describes* how the Spelling and Grammar Checker works, and the after example *prescribes* how to use the Spelling and Grammar checker. For some types of writing, this is a critical skill.

The third set of examples shows how you can use an introductory adverbial clause to be more specific or prescriptive. In the rewrites, the first one is more specific than the original. The second one is more prescriptive than the original. This is one more reason why you should add introductory adverbial clauses to your skillset.

In all of these examples, the improvements don't have much effect on the GL scores. In fact, the GL score for some of the improvements is higher than it was in the original. But don't let that diminish the importance of being specific and prescriptive. In fact, if your writing isn't specific, it isn't likely to be effective. And not being specific is a common problem in business writing.

Be specific

An excerpt that isn't specific (GL=14.9)

Recent changes to the acquisition regulations make clear that the degree of analysis and documentation supporting an acquisition should match the size and complexity of the need. Therefore, one to ten pages of documentation are required to justify the need and obtain the approval for a purchase.

Rewritten with specific information (GL=13.9)

To get approval for a purchase of from $1,000 to $5,000, you need to write a justification of from 1 to 3 pages. To get approval for a purchase of from $5,000 to $50,000, your justification needs to be from 4 to 10 pages.

Be prescriptive, not descriptive

An excerpt that's descriptive (GL=9.5)

The Spelling and Grammar Checker is used to catch the spelling and grammar errors in a document. When this checker is run, a dialog box is displayed whenever an error is detected, and that box requires a response.

Rewritten so it's prescriptive (GL=6.8)

To catch the spelling and grammar errors in a document, you can run the Spelling and Grammar Checker. To do that, click on the Spelling and Grammar icon in the Review tab of the Word ribbon. Then, respond to each of the dialog boxes that are displayed.

Use introductory adverbial clauses to be more specific or prescriptive

Example 1: Not specific (GL=6.7)

Mr. Clark is open and friendly, and I prefer him to the other candidates.

Rewritten so it's more specific (GL=7.5)

Because Mr. Clark is open and friendly, I prefer him to the other candidates.

Example 2: Descriptive (GL=7.3)

Word's Spelling and Grammar Checker is used to check the spelling and get the GL score for a document.

Rewritten so it's prescriptive (GL=8.2)

If you want to check the spelling and get the GL score for a document, you should use Word's Spelling and Grammar Checker.

Description

- For most business writing, generalities don't get the reader anywhere, so you need to be *specific*.
- If you're trying to show how to do something, it's usually better to *prescribe* than *describe*.

Figure 4-7 Be specific and prescriptive

Use active voice

When you use *passive voice*, you say, "The report was written by David." When you use *active voice,* you say, "David wrote the report." This is illustrated by the first group of examples in figure 4-8.

The benefit of active voice is that it shows who did what. And that in turn can make a sentence easier to understand. That's why you should use active voice whenever it improves the clarity of your sentences, especially in the main clauses of your sentences.

This is illustrated by the second group of examples. When the original is rewritten with active voice, it is not only easier to read but also makes more sense. But that doesn't mean you should always use active voice.

For instance, the third group of examples shows how passive voice can be used to put the object of a clause first. In the first example, passive voice puts the emphasis on "readability statistics." In the second example, passive voice helps to provide continuity by starting with the pointer word *this.* This use of passive voice is common in business writing, and it usually doesn't detract from the clarity of a sentence.

You should also be aware that the use of passive voice is less important in subordinate clauses than it is in main clauses. This is illustrated by the fourth group of examples. Here, active voice is used in the main clauses, but passive voice is used in the subordinate clauses. In both cases, converting those clauses from passive voice to active voice probably won't make the sentences any clearer.

If you're an alert reader, you may have noticed how often passive voice is used in this chapter and book. For instance, passive voice is used in the main clauses in the second and third sentences in the preceding paragraph. This just shows that passive voice is commonly used in this type of didactic writing...and rewriting it in the active voice isn't likely to improve it.

To illustrate, here's how that paragraph would look if it used active voice in those main clauses:

> You should also be aware that the use of passive voice is less important in subordinate clauses than it is in main clauses. The fourth group of examples illustrates that. Here, the main clauses use active voice, but the subordinate clauses use passive voice. In both cases, converting those clauses from passive voice to active voice probably won't make the sentences any clearer.

Is this any clearer? I don't think so. But you be the judge because that's the type of judgment that you need to make when you write and edit your own work.

The difference between passive voice and active voice

Passive voice

Mr. Clark's employment *was terminated* today, and the HRD department *was notified.*

Active voice

I fired Mr. Clark today, and *I let* HRD know about it.

Use active voice in the main clause most of the time

An excerpt that uses passive voice (GL=7.6)

Apparently, *problems are associated* with our new product. Every day, *complaints* about it *are received* and *sales are lost*. As a result, this *product isn't being sold* by our salesreps even though the *commission percent has been increased.*

The paragraph rewritten in the active voice (GL=6.1)

Apparently, our *new product has* some problems. Each day, dozens of *customers complain* about it, and *we lose* sales. That's why our *salesreps don't sell* this product even though *we've increased* the commission on it.

Passive voice can be used to change the emphasis of a sentence

Active voice (GL=13.9 and 6.9)

After you run the Spelling and Grammar Checker and respond to all of its dialog boxes, *it displays* the readability statistics for a document.

The occasional use of a short question or sentence fragment can also improve your style. The fourth and fifth *examples illustrate* this.

Passive voice changes the emphasis (GL=13.9 and 7.2)

The *readability statistics* for a document *are displayed* after you run the Spelling and Grammar Checker and respond to all of its dialog boxes.

The occasional use of a short question or sentence fragment can also improve your style. *This is illustrated* by the fourth and fifth examples.

Passive voice in adjective and noun clauses is usually okay

In his 1951 book, called *How to Test Readability*, Rudolf Flesch presented a formula for a Reading Ease score *that is* still *used* today.

Way back in 1955, one study showed only an 8 percent increase in reader comprehension when a *manual was rewritten* from a GL of 16 to a GL of 12.

Description

- *Active voice* puts the emphasis on the subject and verb. *Passive voice* puts the emphasis on the object and doesn't identify the subject. That's why you should try to use active voice in the main clauses of your sentences.
- Passive voice is okay in the main clause when you want to start a sentence with the object instead of the subject, but you should consider both alternatives to see which one reads better before you decide.
- Passive voice is usually okay in noun and adjective clauses.

Figure 4-8 Use active voice

Avoid figurative language, trite language, and analogies

Figure 4-9 shows three types of writing that you should avoid because they will hide your meaning. When you use *figurative language*, you use words in a nonliteral way. For instance, the first example uses "exploding" to refer to eLearning courses, "vacuum" to refer to eLearning developers, and "mushroom" to refer to the vacuum. But then, that vacuum is going to be "punctured" by their training programs. Aside from the author's desperate need for some knowledge of physics, this is just silly. And, no, I didn't make this example up.

The second type of language to avoid is *trite language*. *Trite* means overused and thus lacking in interest or originality. For instance, "feature-rich software," "dynamic industry," and "viable project," are examples of trite language that has long since lost its meaning because we've heard these combinations of words so often. The same with "carefully crafted," "firm foundation," and "dramatic departure."

This is illustrated by the examples in this figure. Note here that trite language often includes figurative language that has been overused, like "game plan," "tailored," and "bring ideas to life." In either case, if you rewrite the trite language, your writing will make more sense.

The third type of language to avoid is the use of analogies. When you use an *analogy*, you try to explain or clarify something by comparing it to something that is familiar. For instance, the last example in this figure tries to teach the reader what a class is when you're using the Java programming language by comparing it to a car. The trouble is that most analogies fail, and this one fails dismally.

In the rewritten example, the presentation is straightforward, and the grade level score is lower. Yes, the concept is still difficult, but it's less difficult than trying to figure out how a Java class and a car are related, which by the way is completely off the point. In practice, it seems that analogies are far more likely to make a concept more difficult to understand than they are to provide any clarity.

Nevertheless, people like to use figurative language and analogies. Otherwise, you wouldn't hear and see so much of it. The problem is that so much figurative language is trite and so little is clever. Or, as George Orwell said, most of this language has "lost all evocative power" and is used just to "save people the trouble of inventing phrases for themselves."

So, if you want to improve your writing, you need to avoid the use of figurative language...especially when it's trite. Instead of trying to be clever, just try to be clear. Or, to paraphrase some traditional wisdom on writing: "Whenever you write something that you think is particularly clever, enjoy it for a moment, and then strike it out."

Avoid figurative language

Original (GL=13.3)

Today, the demand for eLearning courses is *exploding*, but a developer *vacuum* exists and it will *mushroom* in the years ahead. Our company will *puncture* that *vacuum* and aid the eLearning *revolution* by training the eLearning developers.

Improvement (GL=10.9)

Today, the demand for eLearning courses is growing rapidly. However, there aren't enough eLearning developers to make the eLearning products that corporations are looking for. To address that need, the goal of our company is to train eLearning developers.

Avoid trite language (GL=5.2 and 10.0)

Our insurance programs are *tailored to your needs*.

Agile development tools are *carefully crafted* to provide the *firm foundation* that helps people *bring their ideas to life*. As such, they mark a *dramatic departure* from traditional tools.

Avoid analogies

Original (GL=12.5)

Just as you cannot drive an engineering drawing of a car, you cannot "drive" a Java *class*. Just as someone has to build a car from its engineering drawings before you can actually drive a car, you must build an *object* of a Java class before you can get a program to perform the *methods* that the class facilitates.

Improvement (GL=6.7)

When you program in Java, you use *methods* that are provided by Java *classes*. However, before you can use any of those methods, you must create an *object* from the class that you're going to use. That object in turn provides access to the methods.

What George Orwell had to say about figurative language

There exists a "huge dump of worn-out metaphors which have lost all evocative power and are merely used because they save people the trouble of inventing phrases for themselves."

Description

- *Figurative language* is the use of words in a nonliteral sense.
- *Trite language* refers to language that has been so overused that it has lost its meaning, and trite language often includes figurative language.
- An *analogy* tries to explain a process or concept by comparing it to something else, but often the comparison just adds to the confusion.

Figure 4-9 Avoid figurative language, trite language, and analogies

Two ways to improve your style

Writing style refers to the way something is written as distinguished from its content. Frankly, if you do everything that you learn in this book, you will have a clear writing style that your readers will appreciate so you won't need to worry about developing your own style. What follows, though, are some ideas that will make your style even better.

Write with a conversational style

Figure 4-10 presents four more guidelines that will help get your ideas across to your readers. The first one is to write with a *conversational style*. In general, that means that you should write the way you talk. But that doesn't mean your writing should be exactly the way you talk. Instead, your writing should be "improved conversation." That in turn means that your writing will be easier to read because it will use common words and manageable sentence structures.

For some reason, though, most business writers don't have a conversational style. Perhaps they think it's more businesslike to write in a formal and impersonal style. This is illustrated by the first example in this figure. In the rewritten version, the language isn't exactly the way you would talk. But it sure improves the writing style.

The second way to improve your style is to write with a "you" language. When you do that, you turn the emphasis away from yourself and toward your audience. That implies a way of thinking as well as a way of writing. This is illustrated by the second example in this figure. And this too makes your writing more conversational.

The third way to improve your style is to use "I" or "we" whenever it's appropriate. That goes along with using a "you" language, but some writers refuse to do that, even when they're the ones giving an opinion.

The fourth way to be more conversational is to use contractions. This is illustrated by the before and after sentences in the fourth group of examples. Here, the difference is minor but the tone is improved.

Incidentally, when you use contractions with the word *not*, like "don't" for "do not" and "wouldn't" for "would not," you also prevent misreading. That's because skipping over the word *not* is a common reading mistake, so what should be a negative is read as a positive. In contrast, people don't miss the negative when you use the contractions.

The fifth way to improve your style is to use *that* not *which* as the first word in a clause that's required (called a *restrictive clause*). Although you can start restrictive clauses with either the word *which* or *that*, Rudolf Flesch says that you should go on a "which hunt" and replace every *which* that starts a restrictive clause with *that*. Then, to take that one step further, you can delete any *that* that isn't necessary. Here again, the difference is minor but the tone is improved.

Write as though you're talking with someone

Without a conversational style (GL=16.2)

The purpose of this email is to ask where, within your organization, it is appropriate for our company to introduce its capability and services. We are developers of customized training materials and end-user oriented explanatory materials. We have particular expertise in developing practical eLearning materials.

Rewritten with a conversational style (GL=9.2)

Just wondering if you're satisfied with the training packages that you've been using. If not, I want you to know that we develop training and reference materials for companies like yours. And our specialty is custom eLearning packages.

Write with a "you" language

Without a "you" language (GL=15.2)

Whether the instructors are looking for the results of quizzes or how long students spend in particular modules, analytics tools offer a deep-dive picture of how students are faring.

Rewritten with a "you" language (GL=12.6)

Whether *you* want to see how *your* students are doing on quizzes or how long they spend on specific modules, *you* can use the analytics tools to get that information.

Use "I" and "we" whenever that's appropriate (GL=10.3)

I want you to know that *we* develop training and reference materials for companies like yours. And our specialty is custom eLearning packages for on-the-job training.

Use contractions

Without contractions (GL=10.5)

If you are looking for eBooks that provide analytics, you cannot do better than VitalSource eBooks.

Rewritten with contractions (GL=10.7)

If *you're* looking for eBooks that provide analytics, you *can't* do better than VitalSource eBooks.

Use *that* not *which* for restrictive clauses ...and delete unnecessary *thats*

With *which* (GL=4.9)

The products *which* are most in demand are the ones *which* we run out of the most.

Rewritten with one *that* (GL=4.6)

The products *that* are most in demand are the ones we run out of the most.

Description

- To write with a *conversational style* means that you should write in a way that's similar to the way you converse...but the writing should be "improved speech."

Figure 4-10 Write with a conversational style

Use some stylistic devices

Figure 4-11 presents five ways to add a little flair to your writing style. These are minor, even trivial, when compared to everything that you've already learned, but they will improve your style. Although all five should be self-explanatory, here's a quick summary.

People like to read about people doing things, and they like to read quotations. Although business documents don't provide much opportunity for doing either, these techniques work well when you have the chance to use them. This is illustrated by the first two examples in this figure.

The third group of examples shows that starting a sentence with "And," "Or," or "But" is okay if you don't do that too often. That's true no matter what your grade-school or high-school teachers may have told you. In fact, starting a sentence with "And" or "Or" is useful for presenting the last item in a series of complete sentences. And starting a sentence with "But" is an informal way to say "However."

The occasional use of a question or sentence fragment can also enhance your style. This is illustrated by the fourth and fifth examples, and most professional writers use these techniques.

Just remember that what you want the most is a clear writing style, and you should have one if you master all the other skills in this book. Then, as you start to develop your own writing style, you can experiment with some of these devices.

Write about people doing things

Original (GL=12.8)

Apparently, our Des Moines office has discovered the secret of selling our new software package, even though it isn't supposed to be competitive.

Improvement (GL=12.8)

Bill Brisbane in our Des Moines office sold 21 copies of our new software package in just one week, even though everyone else says that our package isn't competitive.

Use quotations

Original (GL=10.2)

The customer said that our offer isn't remotely competitive.

Improvement (GL=8.0)

The customer said, "Your offer isn't even close to being competitive."

Start sentences with "And", "Or", or "But"

Example 1: Original (GL=5.6)

She is intelligent. She is motivated. She is also a born leader.

Improvement (GL=4.6)

She is intelligent. She is motivated. *And* she is a born leader.

Example 2: Original (GL=3.0)

We think we have the best service in the industry. However, you can be the judge.

Improvement (GL=1.5)

We think we have the best service in the industry. *But* you be the judge.

Ask a question

Original (GL=9.4)

I don't think there's any question that eLearning is an opportunity for us.

Improvement (GL=7.3)

Is eLearning an opportunity for us? I don't think there's any question about that.

Use a sentence fragment

Original (GL=9.7)

There are many reasons for our success but I think the main reason is that we try harder than any of our competitors.

Improvement (GL=0.5)

Why have we been doing so well? *Because we try harder.*

Description

- If you do everything that's suggested in this book, you will have a professional writing style. Then, as you develop your own personal style, you can try out some of the techniques in this figure.

Figure 4-11 Use some stylistic devices

Perspective

In this chapter, as in all the chapters in this book, the before examples are taken from books, magazines, and documents in the real world. Yes, I'm sorry to say that the sentences in most business writing are that bad. In fact, it's hard to find examples that are easy to read and understand.

In contrast, if you master the skills in this chapter, you will be able to write sentences that are easy to read and understand. You will also have an effective writing style. Of course, that assumes that you've already mastered the skills in chapters 1, 2, and 3 so you know how to use headings and how to write paragraphs.

Also keep in mind that this chapter presents just the best practices for writing sentences...the ones you need for business writing. But if you like to write, by all means read other books on writing. When I was learning to write, I read a few dozen books on writing, and each one had at least a few ideas that helped me write better.

Although they're dated, my favorite writing books are any of Rudolf Flesch's books. However, the only one that's easily available from Amazon is *The Classic Guide to Better Writing*. If you want to start with the classics, you should also read *The Elements of Style* by William Strunk Jr. and E. B. White. Beyond that, there are dozens of other books that will help you refine your writing, and most of them focus on writing sentences.

Terms

readability
grade level (GL) score
non-restrictive clause
non-restrictive phrase
introductory participial phrase
noun clump
big-word clump
pronoun
pointer
deadwood
simple sentence
independent clause
compound sentence
adverbial clause
clause
subordinate clause
main clause
conjunction
descriptive
prescriptive
active voice
passive voice
figurative language
trite language
analogy
writing style
conversational style
restrictive clause

Summary

- *Readability* refers to how easy it is to read a document. One measure of readability is the *grade level (GL) score*. You can get that score by running Word's Spelling and Grammar Checker.
- Although readability scores don't ensure comprehension, it still makes sense to write at a grade level of 10.5 or below.
- The first step toward simplifying sentences is to divide any sentences that are more than 20 words or two lines of text into two or more sentences.
- To further simplify your sentences, you should avoid *non-restrictive clauses* and *phrases* because they interrupt the main thought of a sentence. You should also avoid *introductory participial phrases* because they don't have subjects.
- To simplify your words and phrases, you should avoid wordy expressions, words with prefixes and suffixes, *noun clumps* and *big-word clumps*, technical jargon and acronyms, and constructions like *he/she*, *and/or*, and *item(s)*.
- To use fewer words, you should use simple verb forms as well as *pronouns* and *pointers*. And you should delete unnecessary adverbs and adjectives as well as other *deadwood*.
- The sentence structures that you should use most of the time are *simple sentences, compound sentences,* and sentences with introductory and concluding *adverbial clauses*.
- To avoid the deadening effect of a series of sentences that start with the subject, you can use introductory words, phrases, and adverbial clauses.
- To make sure you say what you mean, you need to be *specific*. For some types of writing, it's also good to be *prescriptive* instead of *descriptive*.
- To say what you mean, you should use *active voice*, not *passive voice*, in the main clauses of most of your sentences. However, you can use passive voice to reverse the sequence of subject and object whenever that makes sense.
- If you avoid *figurative language*, *trite language*, and *analogies*, your sentences will say what you mean because you will be forced to express your thoughts in a straightforward way.
- To write with a *conversational style*, you can use a "you" language, "I" and "we" whenever appropriate, and contractions. You can also replace the word "which" at the start of *restrictive clauses* with the word "that".
- To improve your *writing style*, you can write about people doing things, use quotations, ask an occasional question, use sentence fragments, and start sentences with *and, or,* or *but.*

5

How to use visuals to enhance your writing

Many business documents don't require the use of visuals like diagrams, charts, and spreadsheets. But some documents can be improved by using visuals, especially documents that present difficult content and concepts. Then, the use of visuals will not only enhance your writing, but also simplify it.

That's why this chapter presents the guidelines and skills that you need for using visuals. For the record, though, I was tempted to drop this chapter from the book because it's mostly common sense. But then, I analyzed some recent white papers, and I realized that common sense isn't that common. In fact, most writers don't use visuals in a way that enhances their writing.

Note, however, that this chapter assumes that you already know how to develop the visuals that you need for the types of writing that you do. So, the focus of this chapter is on making the best use of those visuals

Six guidelines for using visuals

This chapter starts with six guidelines for using visuals that will help you avoid the common errors that business writers make. That starts with planning.

Plan the visuals before you start writing

Figure 5-1 presents a simple method for planning the visuals that you're going to use for a document. To do that, you go through each of the headings and subheadings in the heading plan for the document and ask yourself whether a visual would make it easier to understand the related topic or subtopic. If you think it will, you add a notation for the visual to the heading plan. These notations can be handwritten on a printed heading plan or they can be added to a word processing heading plan.

In the heading plan in this figure, notations have been added for four visuals. The first will be a diagram of how eLearning modules relate to the Learning Management System (LMS). The next two will be screenshots that illustrate two of the components of an eLearning course. And the fourth will be a spreadsheet that summarizes the development costs, sales potential, and profit potential for any eLearning modules that the company might develop.

In this case, there is a one-to-one relationship between the first three visuals and three of the subheadings. The fourth visual applies to a heading, which means that it can relate to more than one of the subheadings that follow. You'll see how that works later on.

Note that you do this planning before you ever create any visuals. That way, you don't waste time creating visuals that you won't use. What's most important is that you envision any visuals that will make it easier for your readers to understand your topics and subtopics.

After you plan the visuals, you develop the visuals that you've planned. At this stage, rough drafts are okay because you'll get a chance to refine them when you write the document. But visuals like screenshots shouldn't require any refinement later on.

As you develop the visuals, you may decide that one or more of your planned visuals won't actually enhance the document. In that case, you remove those visuals from your heading plan. When you finish the rough or finished drafts of the visuals, you're ready to write the first draft of the document.

A heading plan that also plans the visuals

Introduction to eLearning
  What eLearning is
  What eLearning modules are
  eLearning and the Learning Management System (LMS) — 1-diagram
The primary components of eLearning
  Slides — 2-screenshot
  Video
  Quizzes — 3-screenshot
  Simulations
Why most eLearning fails
  The limitations of video
  The limitations of slides
  The missing ingredients for most eLearning courses
Is this an opportunity for us? — 4-spreadsheet
  The development costs
  The sales potential
  The profit potential
My recommendation

Guidelines for planning the visuals

- To plan the visuals, you ask whether a visual could improve the presentation for each heading and subheading. If so, you add an appropriate notation to the heading plan.
- After you plan the visuals, you develop them, but at this point rough drafts are okay. Then, if you decide that a visual isn't going to enhance the document, you remove it from your heading plan.

How the visuals can relate to the headings and subheadings

- One visual for a heading that doesn't have subheadings
- One visual for a single subheading
- One visual for a heading that does have subheadings so the visual can apply to all of its subheadings

Description

- *Visuals* are the screenshots, drawings, diagrams, charts, tables, and more that help readers remember key points, evaluate data, and understand relationships.
- Whenever a visual can make it easier for the readers to understand a topic or idea, it makes sense to use the visual. That way, the topic or idea will not only require less text, but will also be easier to understand.
- One way to plan the visuals for a document is to add notations to the heading plan as shown above. Those can be handwritten on a printed heading plan or entered into the heading plan in your word processor.

Figure 5-1 Plan the visuals before you start writing

If necessary, gather the visuals

If a document requires more than a couple of visuals, it sometimes makes sense to gather them all in a separate word processing document. This is illustrated by figure 5-2. Here, the four visuals that are in the heading plan in figure 5-1 have been gathered into one document. To gather these visuals, I took screenshots of an LMS diagram that I found on the Internet, two slides in an eLearning course, and a portion of a spreadsheet that I had developed.

After you gather the visuals, you can analyze each visual, decide whether it's needed, decide whether it needs to be improved, and so on. In short, gathering the visuals gives you a chance to improve the visuals...before you ever start writing. Then, as you write the document, you can cut the visuals from the document that you've gathered them into and paste them into the document you're writing.

Of course, you don't need to gather the visuals if you're just going to use a visual or two in a document and you already have them or know where they are. Then, you can just cut and paste the visuals from their sources into the document that you're writing.

Incidentally, if you're going to use screenshots as visuals, there are several ways to capture them. On a Windows system, for example, you can use its built-in Snipping Tool. And on a Mac, you can use Mac Grab. But if you need to take a lot of screenshots, we recommend an inexpensive, professional tool like SnagIt from TechSmith. For more information about these and other options, you can Google "capture a screenshot."

The first two pages of a Word document that contains four visuals

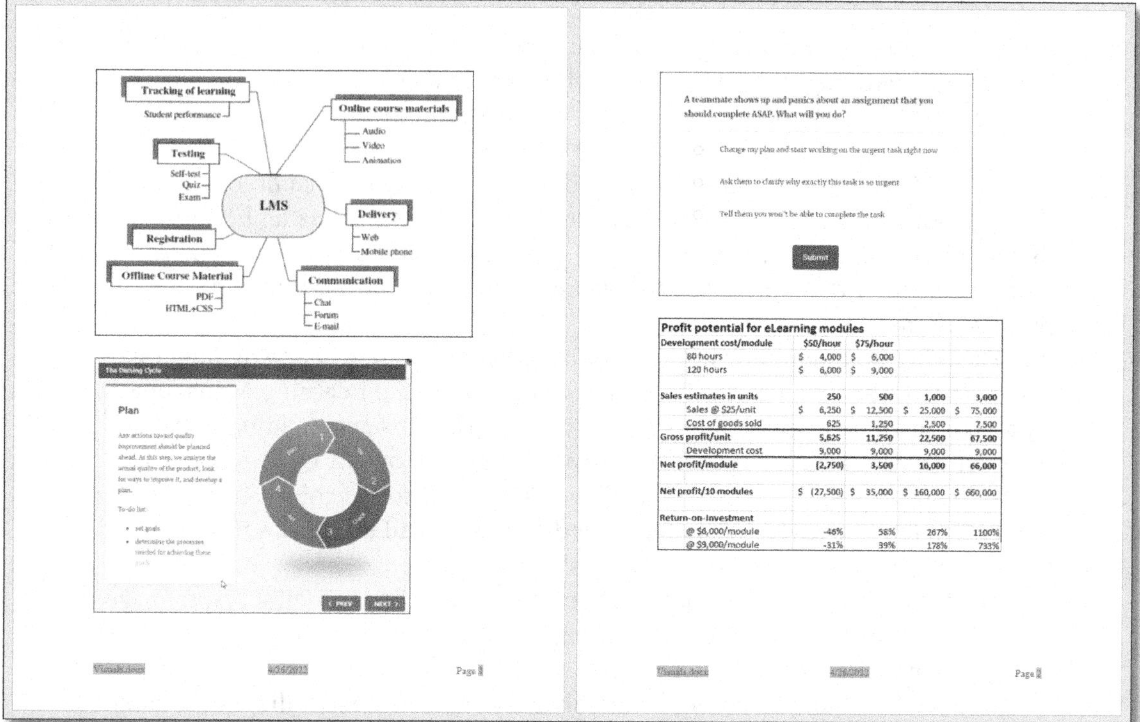

Profit potential for eLearning modules				
Development cost/module	$50/hour	$75/hour		
80 hours	$ 4,000	$ 6,000		
120 hours	$ 6,000	$ 9,000		
Sales estimates in units	250	500	1,000	3,000
Sales @ $25/unit	$ 6,250	$ 12,500	$ 25,000	$ 75,000
Cost of goods sold	625	1,250	2,500	7,500
Gross profit/unit	5,625	11,250	22,500	67,500
Development cost	9,000	9,000	9,000	9,000
Net profit/module	(2,750)	3,500	16,000	66,000
Net profit/10 modules	$ (27,500)	$ 35,000	$ 160,000	$ 660,000
Return-on-investment				
@ $6,000/module	-46%	58%	267%	1100%
@ $9,000/module	-31%	39%	178%	733%

Description

- If you need to create some of the visuals for a document, you should at least develop a rough draft of each of those visuals before you start writing. That will help you decide whether each visual will enhance the document.
- For longer, more complicated documents, you may want to gather the visuals before you start writing. That can help you decide which ones will enhance the text.
- One way to gather the visuals is to paste them into a separate word processing document, as shown above. Later, when you write the document, you can cut and paste the visuals that you've gathered into the document that you're writing.
- To capture screenshots for your documents, you can use built-in tools like the Windows Snipping Tool or Mac Grab. But if you're going to use many screenshots, we recommend a professional tool like TechSmith's SnagIt. For more information about these and other options, you can search online.

Figure 5-2 If necessary, gather the visuals

Be sure that each visual enhances the document

When you plan the figures for a document, you are looking for visuals that enhance the text. For example, a visual of a slide in an eLearning module makes it easier to understand how a slide might work. And a spreadsheet can summarize data in a way that can't be done in text alone.

Often, though, you'll find visuals in business documents that don't enhance the text. Some visuals, like the first one in figure 5-3, are confusing because they're more complicated than they need to be. That often happens when the writer picks up a visual from the Internet even though it doesn't quite fit the purpose of the document.

The other common flaw is to use visuals that don't contribute anything of value. That's illustrated by the second example. Here, the Learning Management System is in an inner circle, and the seven categories that surround it are in an outer circle: Students & Instructors, Master Courses, Content Delivery, and so on. But what relationships does that show? And what do the colors mean? Not to mention that the reverse type makes it difficult to read the text in the ovals. Is this just graphics for the sake of graphics?

The trouble is that any visuals that don't enhance the text detract from it. In fact, they not only waste the reader's time, but also the writer's time. That's why this guideline is so important...even though it's often ignored.

A related problem is that it's relatively easy to make charts, diagrams, and graphs. But it's hard to design them so they will enhance a document. To test that premise, I looked up LMS diagrams on the Internet and quickly found several dozen of them. But not a single one made me say, "Aha, I get it. I see the relationship between the authoring tool and the LMS, I see how they work together, and I see what the LMS does." For that, you would need to create a new diagram of your own.

A visual that's more complicated than it needs to be

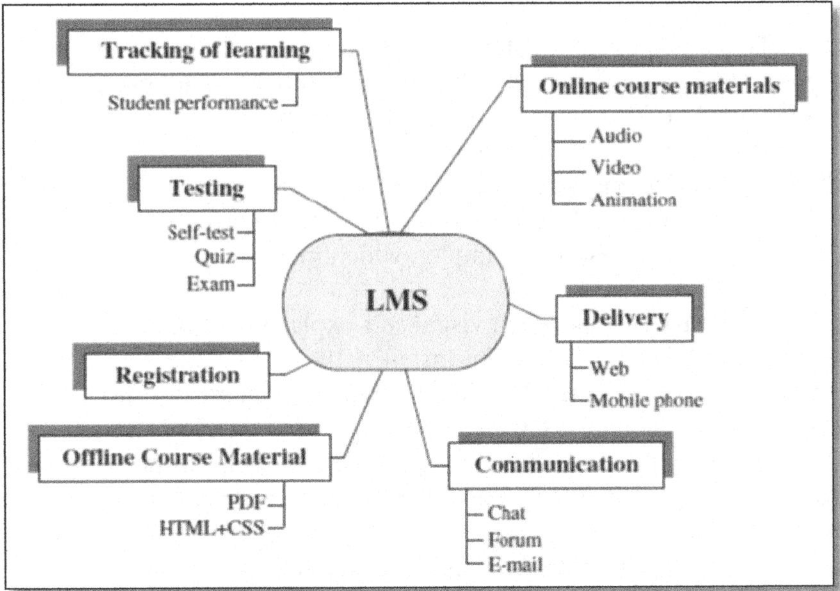

A visual that won't enhance a document

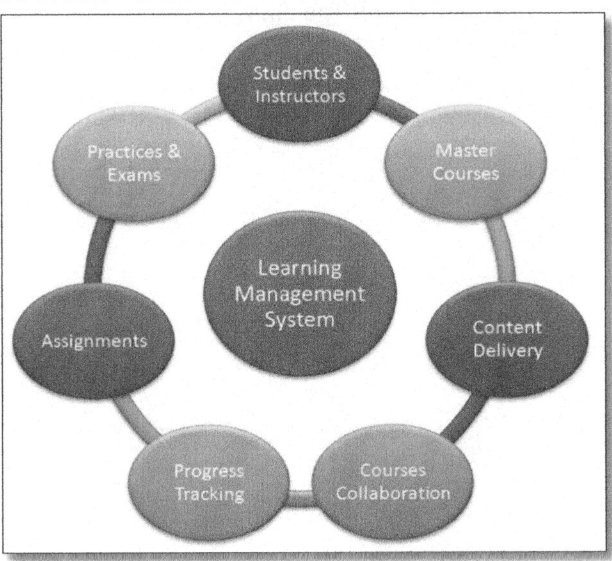

Guideline

- Each visual should make the document easier to understand.

Description

- When a visual *enhances* a document, less text is required and the related topic or subtopic becomes easier to read and understand.
- If a visual doesn't enhance the document, it wastes the time of both the reader and the writer.

Figure 5-3 Be sure that each visual enhances the document

Try to make each visual easy to understand

When you use visuals in a document, the text should explain anything that may be hard to understand. As much as possible, though, you should try to make each visual easy to understand on its own or with minimal explanation in the text. This is summarized by the guidelines in figure 5-4.

The first guideline is to keep it simple. In this figure, both visuals present the same data. In the bar chart, though, the data is difficult, if not impossible, to interpret. In contrast, the line chart is much simpler, which makes it easy to interpret the data.

The second guideline is to try to make each visual self-explanatory. To do that, you can use the title, labels, and callouts. For instance, the title for the second chart in this figure is more specific than the one in the first chart. Besides that, the label "Dollars" has been added at the top of the y-axis for the second chart, so the reader can tell what those numbers represent. And the four groups in the second chart are identified so the reader can tell what they represent without having to read the text.

The third guideline is to make sure that the graphics make the data easier to understand, not harder. In this figure, for example, the graphics in the 3D bar chart look great, but the data is harder to understand. So, yes, it's tempting to create visuals with interesting graphics, but if it obscures the data, that's a temptation you need to resist.

A visual that is difficult to interpret

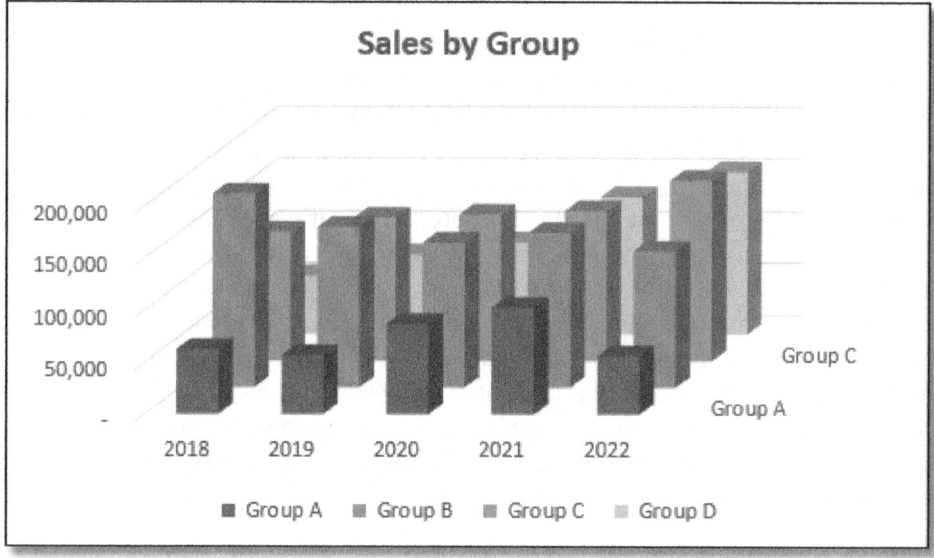

A visual with the same data that's easy to interpret

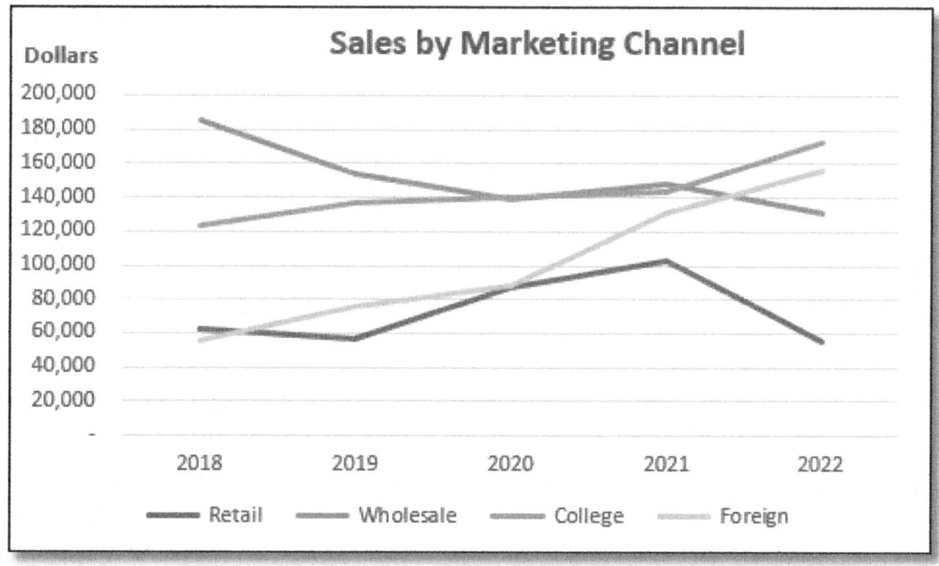

Guidelines

- Keep it simple because the simplest presentation is usually the best.
- Try to make each visual self-explanatory.
- Make sure the graphics make the data easier to understand, not harder.

Description

- As much as possible, each visual should be easy to understand with minimal help from the text.

Figure 5-4 Try to make each visual easy to understand

Be sure that each visual is honest

Look at figure 5-5. Both charts present the same sales data for the years 2013-2022. Here, the first chart indicates that the growth rate is impressive, but the second one indicates a modest growth rate. So, which one is more honest?

If you look at the numbers in both charts, you can see that the sales haven't quite doubled over 10 years, as they went from a little over $800,000 to a little over $1,500,000. That's a growth rate of about 7 percent, and it's even less if you factor in an inflation rate of 3 or 4 percent. That's why the second chart more accurately represents the data.

The difference in the charts is that the x-axis on the first chart starts at $800,000, and the x-axis on the second chart starts at 0. That's one of the many ways that charts can distort data, and that kind of misrepresentation is usually deliberate. But sometimes, a chart misrepresents the data because the data is inaccurate or the chart is charting the wrong data. No matter what the problems are, however, it's up to the writer to make sure that the data is accurate and the visuals are honest.

A visual that misrepresents the data

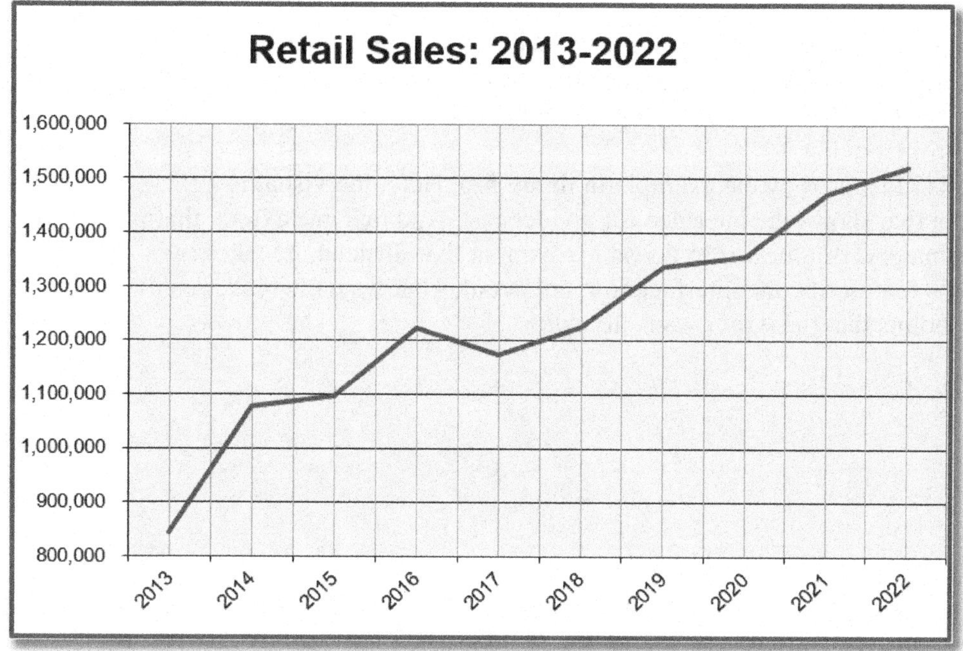

A visual that more accurately represents the data

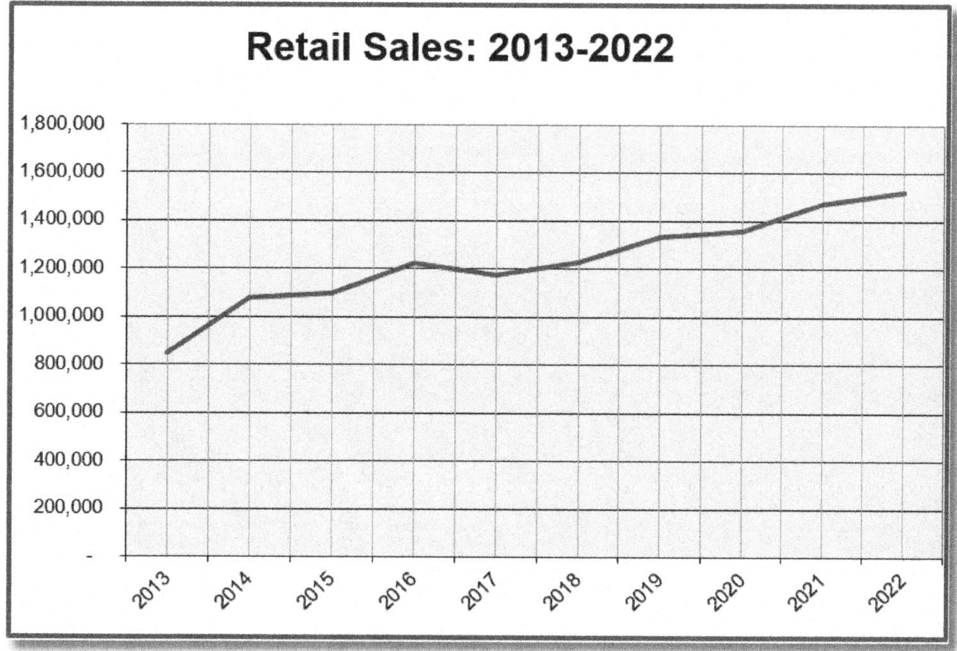

Description

- As much as possible, each visual should accurately represent the data that it presents.

Figure 5-5 Be sure that each visual is honest

Use the text to get the full value from each visual

Another mistake that business writers make is to assume that a visual will get the idea across with just a reference to it from the text. But if you take the time to develop a visual, even if it's just capturing a screenshot, you should try to get the most from it.

This is illustrated by the example in figure 5-6. Here, the visual is a screenshot that shows the interface for a videocast. And to some extent, that is self-explanatory. But the writer doesn't leave it at that. Instead, he takes two paragraphs to describe the interface in more detail. That way, the reader can't miss the points that the writer wants to make.

A document that uses text to get the full value from the visual

Commercial video courses

Two of the largest providers of commercial eLearning courses are Udemy and LinkedIn Learning. Udemy provides courses that range from $10 to $195, and LinkedIn Learning offers all of its courses for a flat monthly fee like $29.99. For these courses, videos and videocasts are the primary delivery systems for the content.

Here, for example, is the Udemy interface for a videocast in a Python course as a portion of code is being described:

In this interface, the left panel shows the videocast, and the right panel shows a transcript of the audio. However, you can close the transcript if you don't want to read it. And you can open a panel on the left that shows the table of contents for the course.

Frankly, the Udemy interface is rather primitive. Here, for example, the coding portion of the screen is so dark that it's difficult to read. In addition, the text in the transcript is divided at points that make it difficult to read. In contrast, LinkedIn Learning has an interface that works better, provides more options, and is easier to use.

Description

- To get the full value from a visual, you need to describe and explain it in the text.

Figure 5-6 Use the text to get the full value from each visual

Two ways to include visuals in your documents

This topic shows you two ways to include the visuals in your documents. This is trivial, but I included it because I remember that I had trouble with this when I first started writing.

Embed the visuals

The first way to include visuals in a document is to embed them in the text. This is illustrated by the example in figure 5-7. Here, a screenshot of a slide is embedded in a paragraph. In this case, the sentence that precedes the visual should end with a colon.

This is the easiest way to integrate visuals with the text. It works best when each visual is only referred to by one paragraph or consecutive paragraphs. But when a visual is referred to from several points in a document, it's better to treat the visuals as separate components. That is illustrated in the next figure.

A document that has an embedded screenshot

Here, for example, is one of the slides for an eLearning course on how to use the Deming Cycle for quality control:

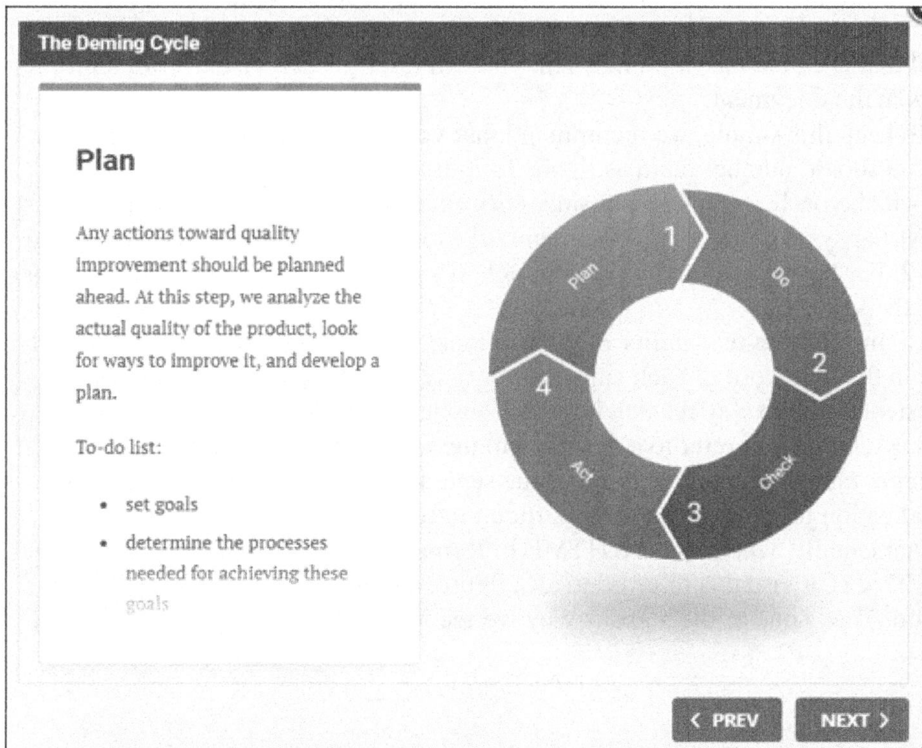

This is like a PowerPoint slide except that the print is usually smaller so it can deliver more text. In addition, it can have an audio track that reads the content that is printed (although a best practice for developing slides says that you should offer either print or audio, not both).

Description

- When you *embed* a visual within a document, you usually introduce it with a sentence that ends with a colon.
- Embedding visuals works best when only one paragraph will refer to each visual. If a visual is going to be referred to from multiple points in a document, it's better to use the technique that's shown in the next figure.

Figure 5-7 How to embed visuals in a document

Treat the visuals as separate components

The other way to present the visuals in your documents is to treat them as separate components. This is illustrated by the example in figure 5-8. In this case, you identify each visual by number and name in a paragraph below the visual that is called the *caption*. Then, you can refer to each visual from multiple points in the document.

To keep this simple, we recommend that you refer to all visuals as *figures*, and you should number them as figure 1, figure 2, and so on. The alternative, which you'll see in some business documents, is to have more than one numbering system in a single document, like chart 1 and chart 2, and table 1 and table 2. For both reader and writer, though, it's easier to use one generic term for all of the visuals.

To improve the readability of the captions, it's best to capitalize just the first letter in the first word of each caption, plus any required capitalization. That is consistent with the way the headings and subheadings should be capitalized. Similarly, when you refer to a figure from the text, you shouldn't capitalize the *f* in *figure* unless it's the first word in the sentence. That's because unnecessary capitalization makes the text more difficult to read.

Incidentally, when you use HTML to format a web page, the same terms are used. That is, a visual is referred to as a figure, and the figure is identified by a caption. That's one of the reasons why we recommend that you use those terms.

A portion of a document that numbers and names the visual

Is this an opportunity for us?

If most eLearning courses fail and the eLearning market is huge, it seems like eLearning might be an opportunity for us. But is it? To answer that, we need to consider the likely costs, sales, and profits of our eLearning courses. This is summarized by the spreadsheet in **figure 4**.

Figure reference

Profit potential for eLearning modules				
Development cost/module	**$50/hour**	**$75/hour**		
80 hours	$ 4,000	$ 6,000		
120 hours	$ 6,000	$ 9,000		
Sales estimates in units	**250**	**500**	**1,000**	**3,000**
Sales @ $25/unit	$ 6,250	$ 12,500	$ 25,000	$ 75,000
Cost of goods sold	625	1,250	2,500	7,500
Gross profit/unit	**5,625**	**11,250**	**22,500**	**67,500**
Development cost	9,000	9,000	9,000	9,000
Net profit/module	**(2,750)**	**3,500**	**16,000**	**66,000**
Net profit/10 modules	$ (27,500)	$ 35,000	$ 160,000	$ 660,000
Return-on-Investment				
@ $6,000/module	-46%	58%	267%	1100%
@ $9,000/module	-31%	39%	178%	733%

Figure 4 Projected costs, sales, and profits for our eLearning courses

Figure caption

The development costs

To develop eLearning courses, we would need to use authoring tools that range from $29.99 per month to flat fees of more than $2,000. So for three developers, let's assume a total fee of $6,000. That one-time cost would of course be manageable if we can build successful products, so this cost isn't included in **figure 4**.

Figure reference

More significant, though, are the variable costs for developing each eLearning course. By our estimates, these costs would range from $4,000 to $9,000 for each course. These fixed and variable costs are summarized in the top portion of **this figure**.

Description

- If you want to refer to the visuals from more than one point in a document, you should number and name the visuals. To do that, you add a paragraph below each visual called a *caption* that provides the visual's number and name.
- To keep it simple, it's best to refer to all visuals with a generic term like *figure*. Then, to make the captions easy to read, it's best to capitalize just the first letter in the first word of each caption (plus any required capitalization).
- Within the text, you shouldn't capitalize the *f* in *figure* unless it's the first word in the sentence.
- In general, if you use this presentation method for one of the visuals, you should use it for all the visuals.

Figure 5-8 When and how to treat visuals as separate components

Perspective

With modern software, it's relatively easy to make visuals like diagrams, charts, and spreadsheets. But surprisingly, many business writers don't use visuals in a way that enhances their writing. So, even though everything in this chapter is common sense, be sure to use your own as you apply it.

Terms

visual
embedded visual
figure
caption

Summary

- After you plan the headings and subheadings for a document, you should plan the *visuals* that will enhance the topics or subtopics in the document.
- After you plan the visuals, you should develop rough or final drafts of them. You may also want to gather them in a separate word processing document. That will give you another chance to analyze and improve them.
- Each visual that you use should (1) enhance the presentation of the text, (2) be as self-explanatory as possible, and (3) be honest. Then, to get the full value from each visual, you need to describe and explain it in the text.
- When you add visuals to a document that you're writing, you can either *embed* them in the text or treat them as separate components.
- When you treat the visuals as separate components, you should refer to the visuals as *figures* and to the paragraph that numbers and names each figure as a *caption*. These are the same terms that are used in the HTML for web pages.

6

How to write a good first draft in record time and edit it without "thrashing"

The goal of this chapter is to show you how to write the first draft of a document in record time, and then edit it without "thrashing." To do that, you'll learn how to treat writing, editing, and proofing as separate processes. When you do that, you will not only write faster, but also better.

This chapter is important because many writers take far too long to write and edit their documents. For example, some write, and then edit, and then edit again...and then edit again. In fact, some books on writing suggest that you work that way...that you should keep editing your work until you get it exactly the way you want it. When you do that, though, the third or fourth edit of the document may not be any better than the second one. And when that happens, you're "thrashing."

As you might guess, the focus of this chapter is on writing lengthy documents like reports and proposals because those are the ones that cause business writers to "thrash." In contrast, it's okay to write a short document like an email, proof it on screen, and send it off. But even then, it's best to separate the three processes: write it, edit it if necessary, and proof it.

How to write the first draft in record time

If you like to write, this is the development step that you've been waiting for. You've already planned what you're going to write and you've developed any visuals that you need. Now, you just sit at your word processor, start writing, and keep going until you're done. And when you're done, you can relax because you know that all your ideas are on paper so all you have to do is edit them.

The first five figures in this chapter present the guidelines that you need for writing a good first draft of a document in record time. In brief, you start the document with the headings and subheadings that are in your heading plan. Then, you write the paragraphs for each of those topics and subtopics, and you insert the visuals into the document wherever they are needed.

Start the document from your heading plan

Figure 6-1 shows a document in Microsoft Word that contains the headings and subheadings for a heading plan. Once you get the heading plan the way you want it, you can write the paragraphs for each topic and subtopic. If you use Microsoft Word, chapter 9 shows you how to use Outline view to create a heading plan and Print Layout view to write the document.

The easiest way to get the headings and subheadings into a document is to use the same document for developing the heading plan that you use for writing the document. In other words, you start a new document, and you enter the heading plan into it. Then, when you've got the heading plan the way you want it, you start writing.

The heading plan for a document in Word's Outline view

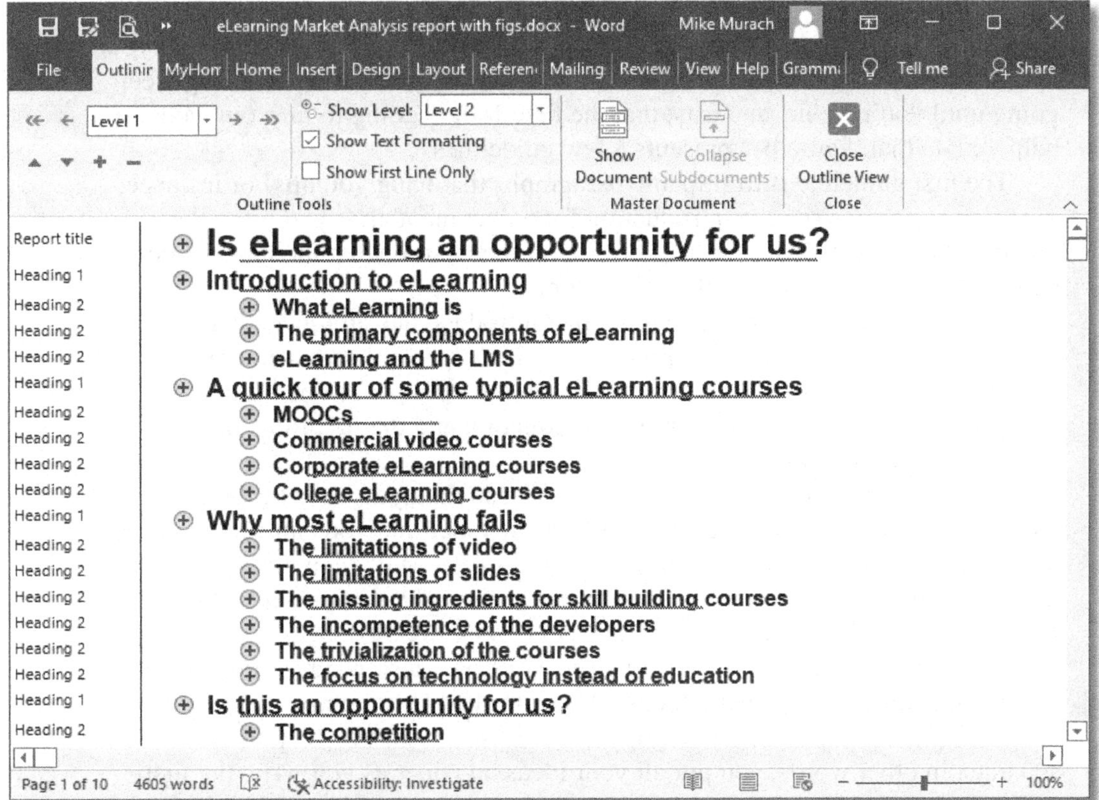

How to use Microsoft Word to start a document from its heading plan

- Use Outline view to develop the heading plan for the document.
- When you've got the heading plan the way you want it, switch to Print Layout view and start writing.

Description

- In chapter 9, you can learn how to use Microsoft Word's Outline view to create and use a heading plan.

Figure 6-1 Start the document from your heading plan

Start somewhere and keep going!

When you write the first draft of a document, your goal should be to get your ideas on paper as quickly as possible. To do that, you need to start writing, keep going until you're done, and trust that the first draft is going to turn out well. To help you do that, figure 6-2 presents a few guidelines.

The first guideline is to skip any paragraphs that hang you up. For instance, it's often easier to write the introduction for a document after you write the document than before you write it. So, if you're not sure what the introduction should do, skip it and start with the first topic.

Similarly, the paragraphs in a topic that introduce the subtopics that follow can be hard to write. Those paragraphs are called *topic openers*, and you can skip them too if they're giving you trouble. In the example in this figure, the writer has skipped both the document introduction and the first topic opener, and has started with the first subtopic.

Once you start writing, just keep going. To do that, you need to:·Stop worrying about how good your writing is. Remember that you can't improve on a blank page. Trust that your heading plan and visuals will keep you focused on the right content. And know that if you write paragraphs that present all the ideas for each topic and subtopic, your first draft is going to be good.

To write the first draft as quickly as possible, you also need to keep your editing to a minimum. To help you do that, you need to remember that writing and editing are two different thought processes, so it's best to keep them separate. In other words, you put all your ideas on paper as you write the first draft. Then, you refine and improve those ideas when you edit the first draft.

After you write the paragraphs for a subtopic, for example, it's okay to review the paragraphs to make sure you've included all of the essential information. Then, if you need to add something, it's okay to do that. But don't try to edit each paragraph so it's perfect because that's what you'll do when you edit the first draft. Instead, keep on writing until the first draft is done.

Later, after you finish the first draft and start to edit it, you're likely to be surprised by how well the first draft turned out. So, keep the faith and keep going, even if you're uncomfortable with that. You'll soon see that this really is the best way to write that first draft.

If they hang you up, skip the introduction and topic openers

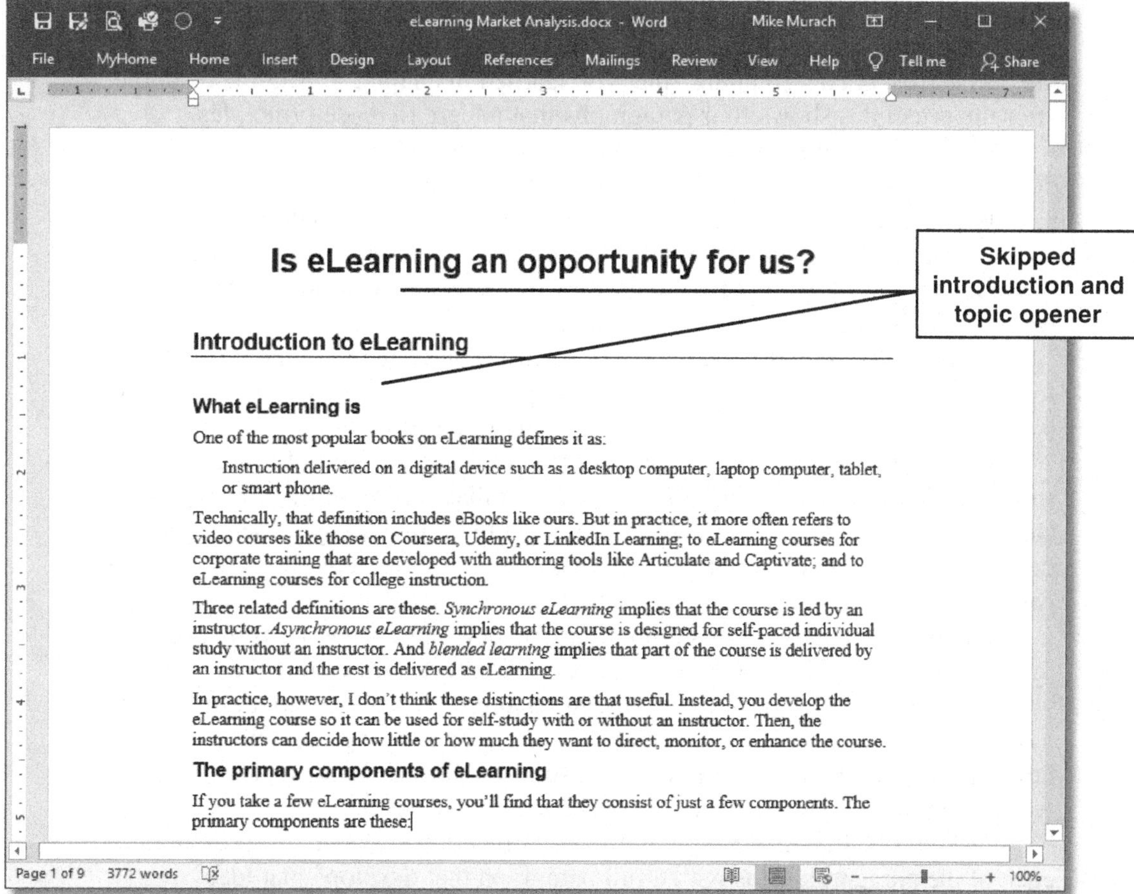

Is eLearning an opportunity for us?

Introduction to eLearning

What eLearning is

One of the most popular books on eLearning defines it as:

> Instruction delivered on a digital device such as a desktop computer, laptop computer, tablet, or smart phone.

Technically, that definition includes eBooks like ours. But in practice, it more often refers to video courses like those on Coursera, Udemy, or LinkedIn Learning; to eLearning courses for corporate training that are developed with authoring tools like Articulate and Captivate; and to eLearning courses for college instruction.

Three related definitions are these. *Synchronous eLearning* implies that the course is led by an instructor. *Asynchronous eLearning* implies that the course is designed for self-paced individual study without an instructor. And *blended learning* implies that part of the course is delivered by an instructor and the rest is delivered as eLearning.

In practice, however, I don't think these distinctions are that useful. Instead, you develop the eLearning course so it can be used for self-study with or without an instructor. Then, the instructors can decide how little or how much they want to direct, monitor, or enhance the course.

The primary components of eLearning

If you take a few eLearning courses, you'll find that they consist of just a few components. The primary components are these:

The goal of the first draft

- To get all your ideas on paper as quickly as possible.

How to start writing

- Don't get hung up on the introduction to the document or *topic openers* (paragraphs that introduce subtopics). Skip those if you have to, but start writing.
- Remember that you can't improve on a blank page, so just get started somewhere.

How to keep going

- Don't worry about how good your writing is because you can fix any problems when you edit the document.
- Trust your heading plan and visuals to keep you focused on the right content. Just try to write one good paragraph after another.
- Keep your editing to a minimum. Remember that writing and editing are two different thought processes, so you do them best when you separate them.

Figure 6-2 Start somewhere and keep going!

Write one good paragraph at a time

Because the paragraph is the primary unit of good writing, figure 6-3 repeats the "principles of paragraphing" that you learned in chapter 3. And the example after the principles shows four paragraphs that adhere to those principles.

One of the benefits of planning the headings and subheadings for a document is that you divide a large writing project into topics and subtopics that usually don't require more than five or six paragraphs. As a result, you should be able to write the series of paragraphs that are needed for each topic or subtopic without any special planning. You just write with a conversational style as if you were presenting the sequence of ideas to a friend.

Occasionally, though, you may have a hard time visualizing the flow of ideas for a topic or subtopic, especially if it's complicated. Then, you may want to step back and plan the paragraphs that you're going to write.

To do that, you just list the ideas that you think are required for the topic or subtopic. You can do that right in the document that you're writing. Then, you can rearrange the paragraph ideas until you get them into a logical sequence for presentation. When you're satisfied that you have a good plan, you can write the paragraphs.

Note in the example in this figure that you don't have to write full sentences for your paragraph ideas. In fact, you can use whatever notation makes sense to you. Here, the notation implies that the first paragraph will talk about how slides have been used for business presentations. The second paragraph will talk about how slides are being used for eLearning. And the third and fourth paragraphs will present the two main criticisms of the use of slides.

Once you feel comfortable with the ideas that you've listed, you can replace each of the paragraph ideas with a full paragraph that develops that idea. Because this is a fluid process, it shouldn't slow you down much as you write that first draft.

The principles of paragraphing

- Start each paragraph with the idea of the paragraph.
- Put one and only one idea in each paragraph.
- Fully develop the idea in each paragraph.

A subtopic with four paragraphs that adhere to those principles

The limitations of slides

For many years, the use of PowerPoint slides for business presentations has been the norm. Slides have also been commonly used for classroom instruction in which the instructor explains what the slides present. Nevertheless, the use of PowerPoint slides has some serious limitations that are loudly criticized. In fact, some companies don't allow the use of PowerPoint for business presentations.

In recent years, slides have also become a primary medium for delivering the content in eLearning courses. Unfortunately, the same criticisms that apply to the use of PowerPoint slides for business presentations also apply to the use of slides in eLearning courses.

The primary criticism is that slides force you to break down the content into chunks that are so small that it's hard for the student to see how the chunks relate to each other. But it's seeing those relationships that lead to deeper understanding and the ability to apply what you've learned. For programming subjects, this means that there's no way to present something like a lengthy program listing without breaking it down into pieces that don't make sense by themselves.

The other criticism of the use of slides for eLearning content is that it forces you to present the content in a rigidly sequential way. To see how the chunks presented by the slides are related, the learner has to go forward through the slides, one at a time. And to review, the learner has to go backwards and then forward through the slides. That's the only way to see the relationships between the slides.

If necessary, plan the paragraphs before you write them

The limitations of slides

PowerPoint slides have long been the norm for business presentations

Now slides have become a primary medium for eLearning courses

One criticism of slides is their sequential nature

The other criticism of slides is that the chunks are too small

Description

- To write well, you need to write one good paragraph after another in a logical sequence.
- For most topics or subtopics, you should be able to write the paragraphs without planning them, just as you would if you were explaining something in a conversation.
- When a topic or subtopic is so complicated that you're having trouble writing its paragraphs, it sometimes helps to step back and plan the paragraphs. To do that, you just list the paragraph ideas, analyze them, and make the necessary adjustments.

Figure 6-3 Write one good paragraph at a time

Insert the visuals as you write

If you have planned and developed the visuals for the document, as shown in chapter 5, you insert them whenever they're needed as you write the text. This is illustrated in figure 6-4. Here, the author has inserted a screenshot of a portion of a spreadsheet into the document at the point where the text first refers to it.

In this case, the spreadsheet is treated as a separate component, and it is referred to by a figure number. The other way to insert a visual into a document is to embed it, as shown in figure 5-7 of chapter 5.

After you insert the visual into the document, you want to be sure that you use the text to get the full value from the visual. To do that, you refer to it, describe it, explain it...do whatever it takes to make sure that the readers understand the purpose of the visual.

As you write the text that's related to a visual, though, you may discover that the visual needs to be adjusted. Then, if it's easy to make the adjustment, you can do that and continue writing. Otherwise, you can just note the adjustment that needs to be made, continue writing, and fix the visual after you've finished the first draft of the text.

What you don't want to do is interrupt the flow of your writing, especially if you're on a roll. If, for example, you want to make some adjustments to the spreadsheet shown in this figure, you can just note the adjustments while you're writing. Then, when you finish the text, you can use Excel to modify the spreadsheet, take a screenshot of it, and paste it into the Word document.

A portion of a document that contains a visual

Is this an opportunity for us?

If most eLearning courses fail and the eLearning market is huge, it seems like eLearning might be an opportunity for us. But is it? To answer that, we need to consider the likely costs, sales, and profits of our eLearning courses. This is summarized by the spreadsheet in figure 4.

Profit potential for eLearning modules					
Development cost/module		**$50/hour**	**$75/hour**		
	80 hours	$ 4,000	$ 6,000		
	120 hours	$ 6,000	$ 9,000		
Sales estimates in units		**250**	**500**	**1,000**	**3,000**
	Sales @ $25/unit	$ 6,250	$ 12,500	$ 25,000	$ 75,000
	Cost of goods sold	625	1,250	2,500	7,500
Gross profit/unit		**5,625**	**11,250**	**22,500**	**67,500**
	Development cost	9,000	9,000	9,000	9,000
Net profit/module		**(2,750)**	**3,500**	**16,000**	**66,000**
Net profit/10 modules		$ (27,500)	$ 35,000	$ 160,000	$ 660,000
Return-on-Investment					
	@ $6,000/module	-46%	58%	267%	1100%
	@ $9,000/module	-31%	39%	178%	733%

Figure 4 Projected costs, sales, and profits for our eLearning courses

The development costs

To develop eLearning courses, we would need to use authoring tools that range from $29.99 per month to flat fees of more than $2,000. So for three developers, let's assume a total fee of $6,000. That one-time cost would of course be manageable if we can build successful products, so this cost isn't included in figure 4.

Description

- As you write the first draft of the document, you insert each visual into the document at the point where it is first needed.
- As you write the text for a visual, you may discover that the visual needs to be modified. But don't let that slow down your writing. Just note the required changes, keep writing, and fix the visual after you finish the first draft of the text.
- Be sure to get the full value from your visuals by using the text to explain and expand upon what the figures present.

Figure 6-4 Insert the visuals as you write

If necessary, adjust the heading plan

As you write the first draft of a document, you may discover that the headings and subheadings aren't working out quite the way that you planned. For instance, you may decide that you don't need to present some of the subtopics in as much detail as you originally envisioned. Or you may decide that the sequence of topics or subtopics needs to be changed.

In cases like that, you just adjust the heading plan as shown in figure 6-5. In this example, the starting heading plan has a topic on the components of eLearning that consists of four subtopics. But then the author decides that (1) the primary components should be treated as part of the introduction to eLearning, and (2) a full subtopic doesn't need to be devoted to each component. These adjustments have been made in the second example in this figure.

To make adjustments like that as you write the text, you can of course cut-and-paste portions of the text. If you're using Microsoft Word, though, it's often easier to use Outline view to make the changes. To do that, you switch to Outline view, make the adjustments, and switch back to Print Layout view.

The good news is that this type of change is easy to make when you use a heading plan to plan what you're going to write. That's because each topic and subtopic is treated as an independent module. As a result, you won't have to make many changes to the text when you move or delete topics and subtopics.

In the example in this figure, you just move the heading for the primary components under the Introduction to eLearning heading, and you convert the components heading to a subheading. Then, you delete the subheading for each component, and you describe each component in a paragraph or two instead of a full subtopic. You probably won't have to make any other changes to the text.

The starting heading plan for an eLearning report

- Introduction to eLearning
  - What eLearning is
  - eLearning and the Learning Management System (LMS)
- The primary components of eLearning
  - Slides
  - Video
  - Quizzes
  - Simulations
- A quick tour of some typical eLearning courses
  - MOOCs
  - Commercial video courses
  - Corporate eLearning courses

The adjusted heading plan

- Introduction to eLearning
  - What eLearning is
  - The primary components of eLearning
  - eLearning and the Learning Management System (LMS)
- A quick tour of some typical eLearning courses
  - MOOCs
  - Commercial video courses
  - Corporate eLearning courses
  - College eLearning courses

When to adjust the heading plan as you write

- Whenever you discover that the topics or subtopics that you've planned aren't working out the way you envisioned.

How to adjust the heading plan as you write

- Cut and paste the topics and subtopics until you get them the way you want them.

How to adjust the heading plan with Microsoft Word's Outline view

- Switch to Outline view, display just two levels of headings, and adjust the headings and subheadings.
- Switch back to Print Layout view, and continue writing.

Description

- When you divide a document into topics and subtopics, the topics and subtopics are relatively independent. That's why it's easy to adjust the heading plan as part of the writing process.
- In chapter 9, you can learn how to use Word's outline feature to develop a heading plan and how to use that feature to adjust the heading plan as you write.

Figure 6-5 If necessary, adjust the heading plan

How to edit without "thrashing"

As you have just learned, when you write the first draft of a document, your goal is to get the required information on paper in a series of fully-developed paragraphs. After that, your goal is to edit the first draft just once, proofread it just once, make any final corrections, and be done. Anything beyond that, we call "thrashing."

To edit short documents, you can just read the document on your screen and make any changes or corrections as you read. That works okay if the document is just a couple of pages and not too important. But if a document is important like a report or proposal, you need to work more methodically. And that's what you'll learn how to do next.

Edit the first draft just once

Figure 6-6 summarizes our guidelines for editing the first draft of a document like a report or a proposal. Most important is that you *print* the document and mark any changes or corrections on the paper. You shouldn't read the document on your screen and make changes on the fly.

The main reason for working with printed documents is that you don't make the changes until you've read and marked the entire document. But by that time, you may realize that changes to one paragraph will affect another paragraph farther down the document. In contrast, when you read a document on screen, you never get that perspective.

As you read and edit, you use mental checklists like the ones in this figure. Does the paragraph I'm reading present one and only one idea? Is the paragraph idea fully developed? Do I need to make the paragraph easier to read and understand?

Then, after you read all the paragraphs for a topic or subtopic, you can pause and take a broader view. Do the paragraphs present all of the required ideas? Do all the paragraph ideas apply to the topic or subtopic? Are the paragraphs in the best presentation sequence. Does each visual enhance the text?

As you read and reflect, you mark the changes on the printed pages. To do that, you can use your own notations. Or, you can use the standard editing marks that are presented in figures 6-12.

Sometimes, though, you will make so many marks for a paragraph that you can't fit any more into the margins. In a case like that, you can just indicate that the paragraph needs to be rewritten. Similarly, if a paragraph presents a fact that you need to check or an issue that needs to be resolved, you can just put a circled question mark next to it to show that you need to check it. That's better than interrupting your editing to check the fact right away.

When you finish marking the editorial changes on the printed first draft of a document, you use your word processor to make those changes. Then, you print the document again and check *just* the changes that you've made to be sure you've done them right.

A document with editorial marks

Is eLearning for us?

During the last few years, eLearning has become a huge business, ~~in the form of~~ websites that offer ~~eLearning~~ courses, in corporate training programs ~~that offer~~ employee ~~training~~, and ~~even~~ college courses. Now, we're wondering whether this is a market that we should get into. This report analyzes this market and our possibilities for success.

Introduction to eLearning

This report starts with a quick introduction to eLearning. ~~This is a summary of what I learned as I did my marketing research.~~ It provides the background for the ~~market~~ analysis.

What eLearning is

One of the most popular books on eLearning defines it as: Instruction delivered on a digital device such as a desktop computer, laptop computer, tablet, or smart phone. That definition includes eBooks like ours, but it also includes video courses and slideshow courses for corporate training.

Editing guidelines for long documents

- Always work from a printed copy of the document with the text single-spaced just as it will be when the document is distributed. Don't read or proofread on screen.
- If you want to rewrite one or more paragraphs or edit them extensively, just note that in the margin. Don't try to mark all of the changes because that gets unwieldy.
- If you want to research an issue raised by a paragraph, just note that in the margin next to the paragraph. Then, you can resolve the issue later on.

Editing checklist for each topic or subtopic

- Does each paragraph start with the idea of the paragraph and fully develop that idea?
- Does each paragraph apply to that topic or subtopic?
- Do the paragraphs present all of the required information?
- Are the paragraphs in the best presentation sequence?
- Does each visual enhance the text?

Description

- Although it's okay to edit short documents on your screen, it's better to print long documents and edit them on paper. That way, you can work more methodically.
- When you're working alone, you can mark the changes in any way that works for you. But if you want to use the standard editing marks, you can refer to figure 6-12.

Figure 6-6 Edit the first draft just once

If necessary, analyze the paragraphs

As you read and mark a document that you're editing, you may find that you have trouble following the logic of the paragraphs for an occasional topic or two...even though you've written them. Then, it's worth taking the time to analyze them, as shown in figure 6-7.

To start your analysis, you just list the idea of each paragraph. You can do that on a separate sheet of paper or in the word processing document itself. Then, you can use the checklist in this figure to analyze the paragraphs.

If you adhere to the principles of paragraphing when you write the first draft, each paragraph should present one fully-developed idea, so that shouldn't be a problem. Instead, you can focus on whether each paragraph applies to the topic or subtopic, whether each paragraph is required, and whether the paragraphs are in the best sequence for presentation.

In the before and after example, the author decided that the first paragraph on how PowerPoint slides were used in the past wasn't necessary so she deleted it. She also decided that the sequence of the two criticisms should be reversed. That's typical of the type of thinking that you do when you analyze the paragraphs. Then, you make the changes to the document, which are usually easy to do.

If you're an alert reader, you've noticed that analyzing the paragraphs is the counterpart to planning the paragraphs as shown in figure 6-3. Of course, you hope that if you plan the paragraphs when you write, you won't have to analyze them when you edit. But for difficult topics or subtopics, you sometimes have to do both. Just remember that what you're after is to get it right, especially the most complicated material.

A paragraph list for one subtopic in an eLearning report

> **The limitations of slides**
>
> PowerPoint slides have long been the norm for business presentations
>
> Now slides have become a primary medium for eLearning courses
>
> One criticism of slides is their sequential nature
>
> The other criticism of slides is that the chunks are too small

The paragraph list after the author has improved it

> **The limitations of slides**
>
> In recent years, slides have become a primary medium for eLearning courses
>
> The primary criticism of slides is that the chunks are too small
>
> Another traditional criticism of slides is their sequential nature

Editing checklist for each paragraph in the list

- Does each paragraph present one and only one idea?
- Does each paragraph apply to that topic or subtopic?
- Do the paragraphs present all of the required information?
- Are the paragraphs in the best presentation sequence?

What you may decide to do after your analysis

- Add or delete one or more paragraphs
- Change the sequence of one or more paragraphs
- Divide paragraphs that contain more than one idea

Description

- If you have trouble following the logic of some of the paragraphs as you read them...even though you've written them...it sometimes helps to analyze them.
- To start your analysis, you list the ideas for the paragraphs. You can do that with a pencil on paper or with a word processor.
- If you have trouble listing the idea for a paragraph, that's an indication that the paragraph needs to be improved.
- After you list the paragraphs, you can analyze them starting with the checklist above. Then, you can decide what changes you want to make.

Figure 6-7 If necessary, analyze the paragraphs

Proofread the edited document just once

When you *proofread*, or *proof*, a document, you are only looking for errors. You are no longer editing. To do that, you read the document one more time, and mark any errors, as shown in figure 6-8. This time, though, there should be only a few marks per page.

What's important here is that you don't mix editing and proofreading. If you feel compelled to edit a paragraph or two, okay. But if you edit too much, you're no longer proofreading.

Of course, it's always tempting to make just one more improvement. But the more you do that, the more counterproductive that becomes. Just remember that for most business writing, "good enough is good enough." So stop editing when you reach that point.

If you can keep your focus on proofing the document and finding the last few errors, not on doing more editing, proofing should be a pleasure. That's because you will realize that your headings and subheadings have a logical sequence and structure. That your paragraphs present all the information that the topics and subtopics require. And that your paragraphs are easy to read and understand. In short, you've written a document that you can be proud of.

A proofed document with a few corrections

Is eLearning for us?

During the last few years, eLearning has become a huge business. You can see it in websites that offer video courses, in corporate training programs, and online college courses. Now, we're wondering whether this is a market that we should get into. That's why this report analyzes this market and our possibilities for success.

Introduction to eLearning

This report starts with an introduction to eLearning. It provides the background for this analysis.

What eLearning is

One of the most popular books on eLearning defines it as:

> Instruction delivered on a digital device such as a desktop computer, laptop computer, tablet, or smart phone.

That definition includes eBooks like ours, but it also includes video ~~courses~~ and slideshow courses ~~for corporate training~~.

Here are some related definitions. *Synchronous eLearning* is eLearning that is led by an instructor. *Asynchronous eLearning* is eLearning that is designed for self-study. And *blended learning* is eLearning that is partially delivered by an instructor with the rest delivered via eLearning. In practice, however, I don't think these distinctions are that useful. Instead, you develop the eLearning course so it can be used for self study with or without an instructor. Then, the instructors can decide how little or how much they want to direct, monitor, or enhance the course.

Proofing guidelines

- When you *proofread* (or just *proof*) a document, you should only be looking for factual and grammatical errors.
- This is not the time to edit or rewrite. At this point, "good enough is good enough."

Description

- After you edit a document, you should take the time to proof it.
- If you've done everything else right, proofing should help you realize that your document not only achieves its goals but is also easy to read and understand. As a result, all that's left to do is make a few minor corrections.

Figure 6-8 Proofread the edited document just once

Use your word processor to do some final checking

For longer documents like reports and proposals, it's worth taking a few minutes after you proof a document to do some final checking with your word processor. In most cases, that will catch another error or two. Figure 6-9 shows how.

First, you can use the find and replace command of your word processor to check for common errors. If, for example, you know that you often misuse *it's* and *its*, you can search for all occurrences of *it's* and replace the ones you've misused with *its*. Then, you can do the reverse, as shown by the first example in this figure. Usually, you can do several searches in just a few minutes, and that will help you catch an error or two.

You can also use the find and replace command to search for consistent use of capitalization, italics, or other language conventions. If, for example, you realize that you have typed "elearning" instead of "eLearning" in a couple of places in the document, you can use this feature to find and fix any other occurrences.

Second, you can use the spelling and grammar checker of your word processor to catch a few more errors. For instance, the checker in the second example in this figure shows that there's no comma after the introductory phrase "In addition" so it suggests that one be added.

Incidentally, if your word processor doesn't have a spelling and grammar checker, you can search the internet for a separate grammar checker. For instance, a program named Grammarly can be used to check the spelling and grammar in emails, word processing documents, and more.

Be forewarned, however: No matter what grammar checker you use, it will identify errors that aren't really errors. In fact, a significant portion of the errors that it identifies may not really be errors. But that means that you need to know what the correct grammar is so you don't make changes that aren't correct...there are no shortcuts for that.

Use your find and replace command to find and fix common errors

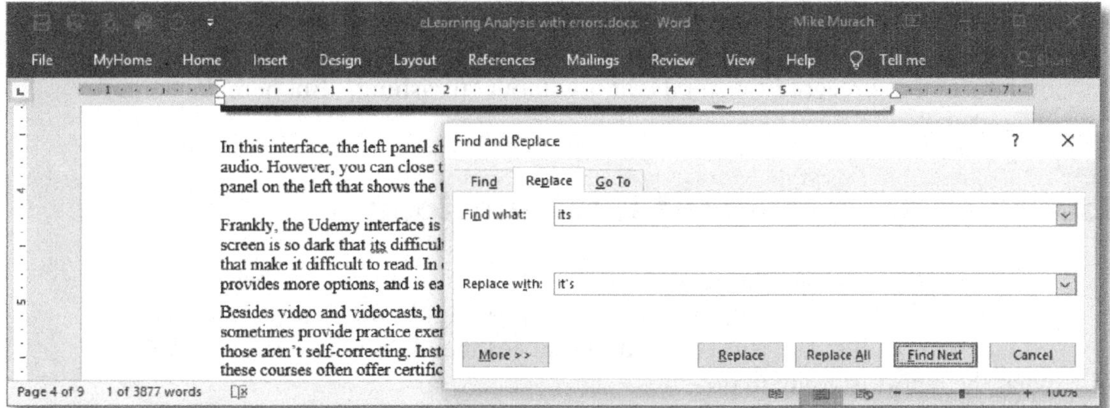

Use your spelling and grammar checker to find any remaining errors

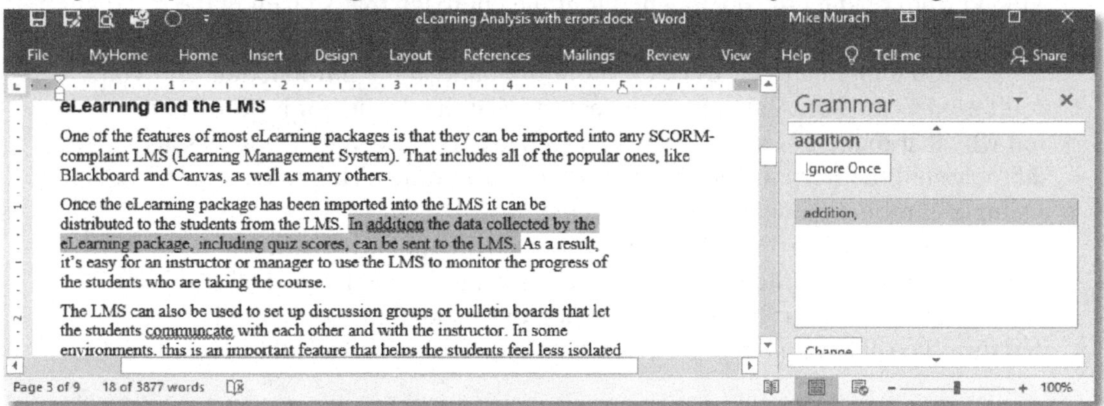

Two ways to use your word processor to do some final checking

- Use the find and replace command to find and fix some common errors.
- Use the spelling and grammar checker.

Typical uses of the find and replace command

- Search for the types of errors that you often make, like the use of *its* when it should be *it's* or the use of *you're* when it should be *your*.
- Search for consistent capitalization, like the use of *elearning* when it should be *eLearning*.

Description

- After you proof a document, you can use the find and replace command of your word processor to find and fix a few more errors.
- You can also use the spelling and grammar checker of your word processor to find and fix errors.

Figure 6-9 Use your word processor to do some final checking

Related skills

The methods that you just learned should help you write a good first draft of a document in record time and edit it without "thrashing." But here are some related skills.

How to write document introductions

As the first example in figure 6-10 shows, the introduction comes right after the title of the document and before the first heading. In this case, the introduction consists of just one paragraph. It introduces the eLearning market and tells what the report is going to do.

In some cases, though, you're going to want to write longer introductions. For instance, you may want to be more specific about the benefits that the reader will get from reading the document. Or, if the approach that's taken in the document is unique, the introduction might also explain that approach.

It's also tempting to try to be clever or dramatic in the introduction. But you don't need to do that. A simple statement of what the document is going to do and why that matters is usually all that you need. Then, you let the topics and subtopics tell the story. In fact, many writers waste time trying to be clever when a simple introduction will be more effective.

As I mentioned earlier, it's often easier to write the introduction to a document after you write the rest of the document. So you may want to skip the introduction when you start writing the first draft and come back to it later. By that time, you'll have a better idea of what that introduction should do.

How to write topic openers

A *topic opener* is usually just one or two paragraphs that come after a heading and before the first subheading for the topic. Once you get used to writing these openers, they shouldn't slow you down at all.

In figure 6-10, each example presents a typical topic opener. Both are short, but they can be as long as they need to be. What a topic opener should do is introduce the subtopics and provide information that's related to all subtopics. What a topic opener shouldn't do is list or introduce each of the subtopics. Because the subheadings make it easy to tell what the subtopics are, listing them in the topic opener just wastes the reader's time.

As I mentioned earlier, if you're not quite sure what a specific topic opener should do, you can write the text for the subtopics and then come back to the opener later on. It's also okay to write topic openers that consist of just one short paragraph, even if it's less than three sentences long.

An introduction and a topic opener in an eLearning report

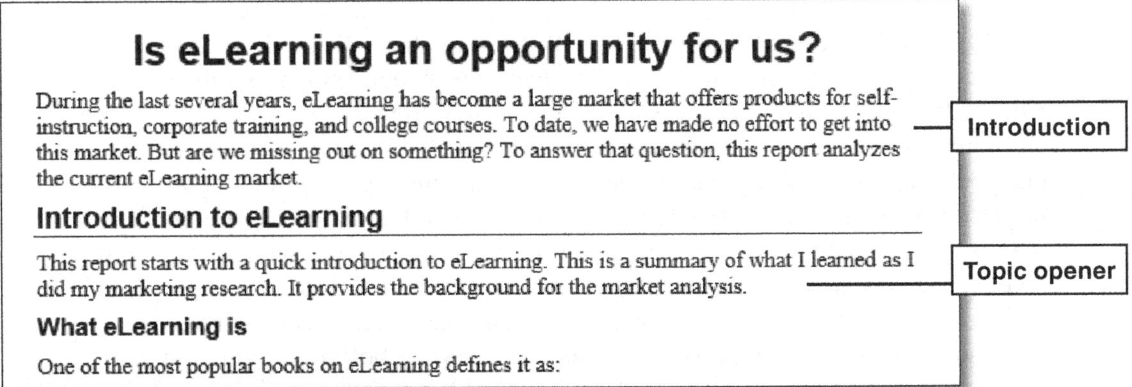

Is eLearning an opportunity for us?

During the last several years, eLearning has become a large market that offers products for self-instruction, corporate training, and college courses. To date, we have made no effort to get into this market. But are we missing out on something? To answer that question, this report analyzes the current eLearning market.

Introduction to eLearning

This report starts with a quick introduction to eLearning. This is a summary of what I learned as I did my marketing research. It provides the background for the market analysis.

What eLearning is

One of the most popular books on eLearning defines it as:

Another topic opener in an eLearning report

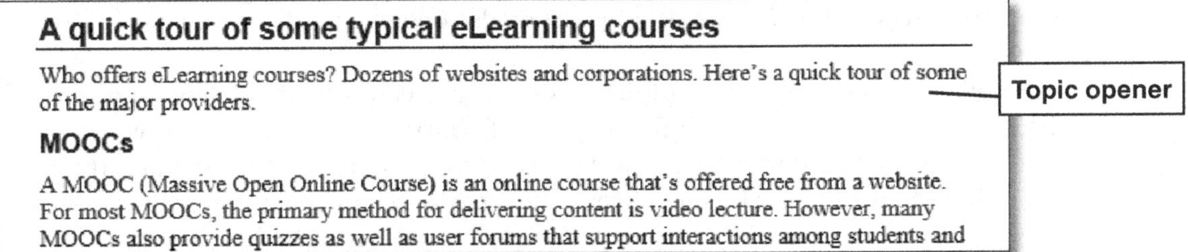

A quick tour of some typical eLearning courses

Who offers eLearning courses? Dozens of websites and corporations. Here's a quick tour of some of the major providers.

MOOCs

A MOOC (Massive Open Online Course) is an online course that's offered free from a website. For most MOOCs, the primary method for delivering content is video lecture. However, many MOOCs also provide quizzes as well as user forums that support interactions among students and

What a document introduction should do

- Introduce the subject and describe the purpose of the document.
- If they aren't obvious, describe the benefits that the reader will get from reading the document.
- If it might be confusing, describe the approach that the document takes to the subject.

What a topic opener should do

- Introduce the topic.
- Provide information that's common to all the subtopics.

What a topic opener shouldn't do

- Introduce each of the subtopics because the subheadings do that.

Description

- The introduction for a document should require just a few paragraphs, and it doesn't need to be clever or dramatic.
- A *topic opener* introduces the subtopics that follow. For many topics, that will require just one short paragraph.

Figure 6-10 How to write document introductions and topic openers

How to format the headings and subheadings

After you've taken the time to carefully plan the headings and subheadings for a document, you want to be sure that your readers can tell them apart. Curiously, though, many business documents don't do that.

This is illustrated by the first example in figure 6-11. Here, the only difference between the headings and subheadings is that the headings are slightly larger. That's okay for the first page of the document, but that doesn't work if there's just one subheading on a page of text. Then, the reader may not recognize whether it's a heading or a subheading, and that difference matters.

In contrast, the second example in this figure has headings and subheadings that are easy to identify. Here, the headings have a bottom border, and the subheadings don't. So, once the readers see those headings on the first page of the document, they'll be able to identify them on subsequent pages too, even if there's only one subheading on a page. That's important because the distinction between headings and subheadings helps the reader understand the structure of the content.

This figure summarizes some of the ways that you can format headings and subheadings so you can easily tell them apart. Of course, what's best is to start each document from a template that has the styles that provide that formatting. If you're using Microsoft Word, you can learn how to do that in chapter 8.

A document with headings and subheadings that are hard to tell apart

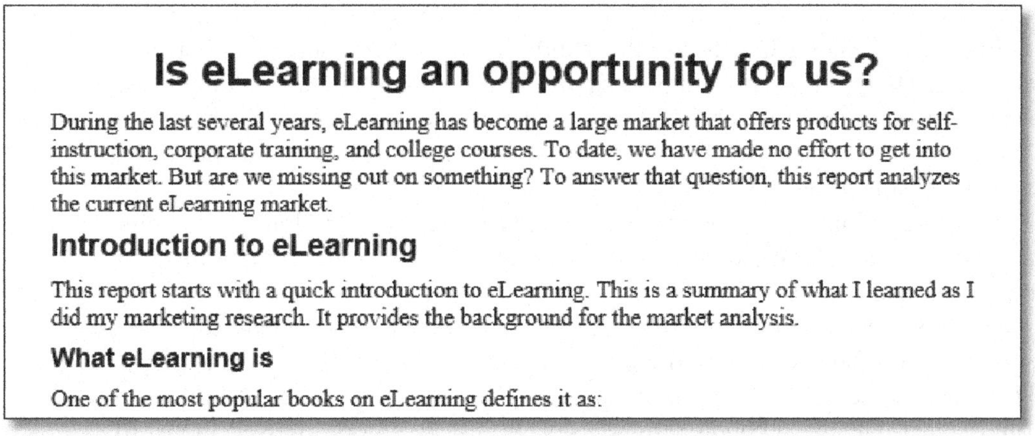

> **Is eLearning an opportunity for us?**
>
> During the last several years, eLearning has become a large market that offers products for self-instruction, corporate training, and college courses. To date, we have made no effort to get into this market. But are we missing out on something? To answer that question, this report analyzes the current eLearning market.
>
> **Introduction to eLearning**
>
> This report starts with a quick introduction to eLearning. This is a summary of what I learned as I did my marketing research. It provides the background for the market analysis.
>
> **What eLearning is**
>
> One of the most popular books on eLearning defines it as:

A document with headings and subheadings that are easy to tell apart

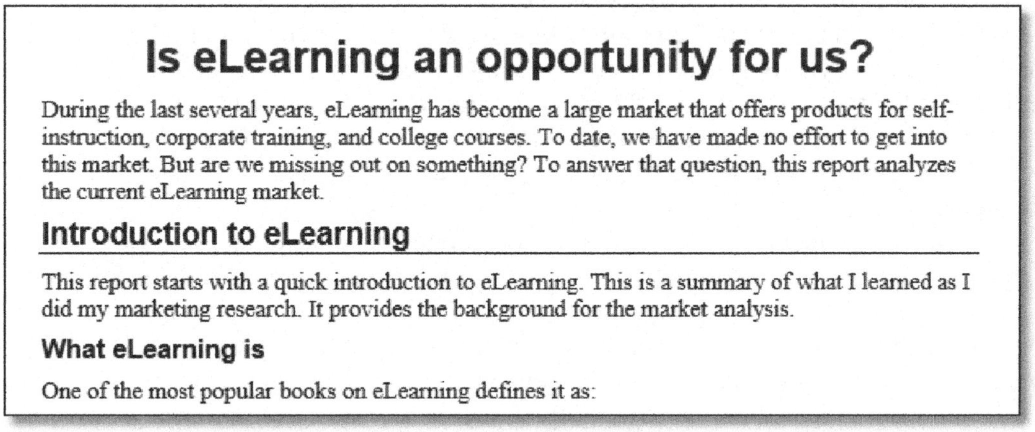

> **Is eLearning an opportunity for us?**
>
> During the last several years, eLearning has become a large market that offers products for self-instruction, corporate training, and college courses. To date, we have made no effort to get into this market. But are we missing out on something? To answer that question, this report analyzes the current eLearning market.
>
> **Introduction to eLearning**
>
> This report starts with a quick introduction to eLearning. This is a summary of what I learned as I did my marketing research. It provides the background for the market analysis.
>
> **What eLearning is**
>
> One of the most popular books on eLearning defines it as:

5 ways to format headings and subheadings so you can tell them apart

- Make the headings larger than the subheadings.
- Put more space before the headings than the subheadings.
- Center the headings and left align the subheadings.
- Add a bottom border to the headings, but not to the subheadings.
- Use a color for the headings, but not for the subheadings.

Description

- After you take the time to plan the headings and subheadings for a document, be sure to format them so the readers can easily tell them apart, even if there's just one on a page.
- The best way to make sure your readers can tell your headings from your subheadings is to start your documents from templates that provide the styles that you need for headings and subheadings. See chapter 8 to learn how that works with Microsoft Word.

Figure 6-11 How to format the headings and subheadings

How to use the standard proofreading marks

Figure 6-12 shows how to use the standard *proofreading marks* as you edit a document. These marks, also known as *proofreader's marks*, have a long history in the publishing industry, and they provide an efficient way to mark changes on a document. They are especially useful when you mark changes that are going to be made by someone else.

If you're going to make your own changes, of course, you don't need to use these marks. But you might want to try the ones that are more efficient than the ones you're using now.

When you use these marks, you first mark the text to show where the change is going to be. Then, you mark the margin to indicate what the change is going to be. For instance, the first example uses the delete symbol to delete four words from the text, and the second example uses the insert symbol to insert the word *just* into the text. Then, in the third example, the transpose symbol is used to switch the order of two words and also two letters.

This figure also presents the proofreading symbols for punctuation. For instance, the first symbol is used for one space; the second symbol is used for the period; and the third symbol is used for the comma. These symbols are used so it's easier to distinguish one handwritten symbol from another. You can see how two of these symbols can be used in the last two examples in the first table in this figure.

To see a larger example of how these marks are used, you can refer back to figures 6-6 and 6-8. Here again, the marks in the text indicate where something is to be done, and the marks in the margins indicate what is to be done. Note that the marks in the margins are applied from the left margin to the right margin, and from left to right when there's more than one mark in a margin.

If you want to learn more about proofreading marks, you can search the internet for "proofreading marks." There, you'll find one-page summaries, articles, and videos on how best to use these marks. But for most business writing, this chapter presents as much as you need to know.

The basic proofreading marks

Symbol	Extension	Meaning	Example	
		Delete	This is an example. It should be self-explanatory.	
^		Insert	This is an example.	just
	tr	Transpose	This only is an example. It should be self-explanatory.	tr tr
≡	cap	Capitalize	This is an example. it should be self-explanatory.	cap
/	lc	Lowercase	This is an Example. It should be self-explanatory.	lc
¶		Paragraph	This is an example. It should be self-explanatory.	¶
	run-on	No paragraph	This is an example. It should be self-explanatory.	run-on
—	ital	Italicize	This is a critical example.	ital
	bf	Boldface	This is a critical example.	bf
		Close up	This is self- explanatory.	
		Delete and close up	This is an exampple.	
		Insert space	This is anexample that should be self-explanatory.	#
		Insert punctuation	This is Johns example.	

The proofreading symbols for punctuation

Symbol	Special character	Symbol	Special character
#	Space		Semicolon
	Period		Apostrophe
	Comma		Quotation mark
	Colon		

The basic procedure for marking the changes

- Use the basic proofreading marks within the text to show *where* in a line a change should be made. Then, mark the change in the left or right margin of that line.
- If there are two or more marks in one margin for a single line, use slashes to separate them.
- The changes are applied from left to right, starting in the left margin and going on to the right margin.
- If the changes that you're marking for a paragraph get unwieldy, stop marking and make a notation that indicates that you're going to rewrite the paragraph.

Figure 6-12 How to use the standard proofreading marks

Perspective

Now that you've read this chapter, you should be able to write a good first draft of a document in record time. When you finish that draft, all your ideas will be on paper, so you just need to edit them. Then, when you edit that draft, your goal should be to improve the writing and correct all errors, but to do that without "thrashing."

As you write and edit, remember that one of the keys to productivity is to keep the writing, editing, and proofreading processes separate. Don't edit when you're writing the first draft. Don't write when you're editing the first draft. And don't edit when you're proofreading. If you can do that, you're on your way to writing like a professional.

Terms

topic opener
proofread
proofreading marks
proofreader's marks

Summary

- When you write the first draft of a document, your goal should be to get all of the ideas for the topics and subtopics on paper as quickly as possible.
- To get off to a fast start, you start each document with the headings and subheadings in your heading plan. Then, you write the paragraphs for each topic and subtopic.
- To start writing and keep going, you may want to skip the introduction to the document or some of the *topic openers* and come back to them later.
- As you write the first draft, you may realize that you need to plan the paragraphs for some topics or subtopics or that you need to adjust the heading plan.
- When you edit a document, you find and fix any errors and you improve the readability of the topics and subtopics. Your goal should be to edit the document just once.
- When you edit a document, you may find that you're having trouble following the logic of the paragraphs for a topic or subtopic...even though you've written them. Then, it's worth taking the time to analyze those paragraphs.
- When you *proofread*, or *proof*, a document, you read the document and try to find any errors that remain. You also resist the urge to do more editing. At some point, you have to say that good enough is good enough.
- After you proofread and fix any remaining errors, you can use the find and replace command of your word processor to do some final checking. You can also use its spelling and grammar checker.
- To get the most from your headings and subheadings, you need to format them in a way that makes it easy for the readers to tell them apart, especially when there's only one heading or subheading on a page.
- To mark errors on the printed pages as you edit and proofread, you can use the standard *proofreading marks*. They provide an efficient way to mark the changes.

7

When and how to use AI writing tools like ChatGPT

Today, you can access many writing tools that use Artificial Intelligence, or AI, to help you write faster. In fact, some studies show that a chatbot like ChatGPT can help you reduce your research and writing time by 30% or more.

That's why you should find out what AI writing tools like chatbots can do and how they might help you. So, after this chapter introduces you to these tools, it shows how to use ChatGPT. When you finish this chapter, you should be able to decide whether you want to use a chatbot for your own research and writing.

Introduction to AI writing and ChatGPT

Artificial Intelligence, or AI, has been used for some types of writing for many years. In the last few years, these writing tools have improved dramatically. And now *ChatGPT*, which was introduced in November of 2022, has taken AI writing to a new level.

An introduction to AI writing

In figure 7-1, you can see the user interface for an AI writing program called Rytr. In the left panel, you can see the request that the writer has made. It consists of the language to be used (English), the tone (convincing), the use case (product description), and the product name (*Murach's HTML and CSS*, which is one of our book titles).

This is followed by a list of features that describes the product: Best-selling book on how to develop web sites; Works for beginning and advanced web developers; Used by more than 200 colleges and universities; and so on. After that, you can set the number of variants that you want produced (2 variants) and the creativity level (Optimal).

Then, when you click the Ryte For Me button, the program pauses for just a few seconds before it displays the response. Because it's hard to read the responses in this figure, here's what the second response (variant) says:

> Get the most out of web development with Murach's HTML and CSS! This best-selling book is an invaluable guide that is trusted by both beginner and advanced web developers alike. With over 200 colleges and universities using this book, you can be sure that you're getting the best instruction available. Don't waste time trying to teach yourself - get Murach's HTML and CSS today and see your sites come alive!

That will give you some idea of how AI writing works, and I would say that this response is darn good...although a professional copywriter would do even better.

As you can see in the first bulleted list in this figure, AI writing tools range from spelling and grammar checkers like Microsoft Word's (see chapter 10)...to more advanced checkers like Grammarly...to special-purpose programs like Rytr and Jasper...to general-purpose programs called *chatbots*. In fact, dozens of AI writing programs are currently available with many more in development.

Today, AI writing is widely used by news organizations like the Associated Press, *New York Times*, and *Wall Street Journal* to prepare sports articles and earnings reports. It is used for creating content for websites that is optimized for search engines (called *Search Engine Optimization*, or *SEO*). And it is used for writing emails and other forms of customer service. But those are just a few of the use cases.

A Rytr prompt and response

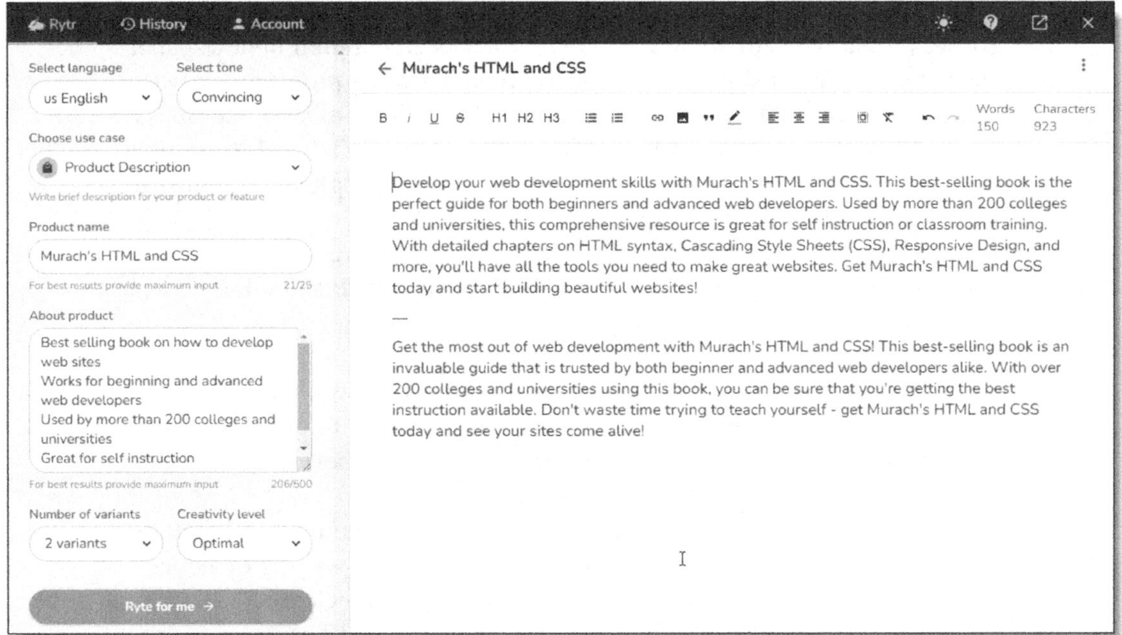

A few of the AI writing tools

- Microsoft Word's Spelling and Grammar Checker (see chapter 10)
- Grammarly, a spelling, grammar, and style checker
- Jasper and Rytr, multi-purpose tools with a focus on blogs and marketing
- Chatbots like ChatGPT, Google Bard, and Windows Bing

Some common uses of AI writing tools

- Sports articles and earnings reports
- Content creation for websites
- Email responses

Some of the users of AI writing

- The Associated Press
- The New York Times
- The Wall Street Journal

Description

- *AI writing tools* range from grammar checkers to tools for specific types of writing to general-purpose tools called *chatbots*.

Figure 7-1 An introduction to AI writing

How ChatGPT has taken AI writing to a new level

You've probably heard about *ChatGPT*. It has been featured in newspaper and magazine articles, and it has been featured on TV news programs. It was introduced by OpenAI on November 30, 2022, and by December 4 had one million users. By February 4, Forbes estimated that ChatGPT had 100 million monthly users, which made it the fasted growing internet application in history. ChatGPT is in a class of AI writing tools called *chatbots*.

To show you what ChatGPT can do, figure 7-2 presents a simple example. Here, the user enters a *prompt* that asks: "What role did NLP and GPT play in the development of chatbots". Then, just a second or two later, ChatGPT starts writing the *response*.

If you read the response, you can get a better idea of how ChatGPT works. That is, ChatGPT uses NLP (Natural Language Processing) to process the prompts, and it uses GPT (Generative Pre-trained Transformer) to generate the results.

The response also says that "GPT models are pre-trained on vast amounts of text data." But that raises questions like: Can ChatGPT's writing be better than the writing in those vast amounts of text data? And can ChatGPT's writing provide information that isn't in those vast amounts of data? Those questions will be addressed throughout this chapter.

Of course, you don't have to understand the underlying technology when you use ChatGPT. Instead, you can treat ChatGPT as a "black box." That is, you give it prompts, and you judge how well it works by the responses. That's the approach taken in the rest of this chapter. That's the approach we recommend for writers. And that's the approach that will show you why you need to master the writing skills that are presented in this book if you want to write like a pro.

Today, chatbots are based on GPT-4, but new and better GPT versions will follow. So you can expect chatbots to get better and better. The bad news is that chatbots are already doing writing jobs that have displaced thousands of workers. The good news is that chatbots may be able to help you research and write faster and better than ever.

A ChatGPT prompt and the start of its response

What role did NLP and GPT play in the development of chatbots?

Natural Language Processing (NLP) and GPT (Generative Pre-trained Transformer) have played pivotal roles in the development of chatbots, empowering them with the ability to understand and generate human-like text. Here's how NLP and GPT have influenced chatbot development:

1. Natural Language Processing (NLP):
   NLP is a subfield of AI that focuses on enabling computers to understand, interpret, and generate human language. NLP techniques form the foundation for chatbot development, providing tools and methodologies to process and analyze textual data. Key NLP components and techniques used in chatbots include:

- Tokenization: Breaking down text into smaller units (tokens) like words or characters.
- Part-of-Speech Tagging: Identifying the grammatical category of each word (noun, verb, adjective, etc.).
- Named Entity Recognition (NER): Identifying and categorizing named entities (person names, locations, dates, etc.) in text.
- Sentiment Analysis: Determining the sentiment or emotion expressed in a piece of text.
- Intent Recognition: Understanding the user's intention or purpose behind a text input.

NLP techniques allow chatbots to process and comprehend user queries, extract relevant information, and generate appropriate responses.

2. GPT (Generative Pre-trained Transformer):
   GPT, based on the transformer architecture, has had a significant impact on chatbot development. GPT models are pre-trained on vast amounts of text data using unsupervised learning, enabling them to learn the statistical patterns and structures of human language. GPT models have several notable contributions to chatbots:

Description

- *ChatGPT* is a general-purpose writing tool called a *chatbot* that lets a user enter a *prompt*. Then, within seconds, ChatGPT starts writing the *response* to the prompt.
- ChatGPT is trained on thousands of pages of writing taken from the internet, books, and other sources. As a result, its writing is as good (or as bad) as the writing in those pages.

Figure 7-2 How ChatGPT has taken AI writing to a new level

Two competing chatbots: Google Bard and Microsoft Bing

Because ChatGPT was first, it is used for most of the examples in this chapter. However, you should also be aware of two competing chatbots, Google Bard and Microsoft Bing, because you may prefer to use one of them for your own research and writing.

These chatbots are illustrated in figure 7-3, which shows the starts of their responses to the same prompt that's used in figure 7-2. That is, "What role did NLP and GPT play in the development of chatbots." If you compare all three responses, you may agree that both the Bard and Bing responses are more to the point than the ChatGPT response. Of course, that's going to vary from one prompt to another, so what's true for one prompt may not be true for another prompt.

Not shown in this figure is that the Bing response provides footnotes for some of the information. Then, you can click on the footnote to go to the web page that is the source of the information. That's a powerful feature. Not to mention that the Bing response seems to be the best of the three.

The start of the Bard response to the question in figure 7-2 : What role did NLP and GPT play in the development of chatbots?:

NLP (Natural Language Processing) and GPT (Generative Pre-trained Transformer) have played a significant role in the development of chatbots.

NLP is a field of computer science that deals with the interaction between computers and human (natural) languages. It is used to process and analyze text, and to generate human-like text.

GPT is a large language model that was developed by OpenAI. It is trained on a massive dataset of text and code, and can be used to generate text, translate languages, write different kinds of creative content, and answer your questions in an informative way.

Chatbots are computer programs that can simulate conversation with human users. They are often used in customer service applications, where they can answer questions, provide support, and resolve issues.

NLP and GPT have been used to develop chatbots that are more natural and engaging than previous chatbots. NLP is used to help chatbots understand the complexities of human language, while GPT is used to help chatbots generate text that is both coherent and human-like.

The start of the Bing response to the same question

Chatbots are software applications that can interact with humans using natural language. They can be used for various purposes, such as customer service, entertainment, education, or information retrieval. NLP and GPT are two technologies that play important roles in the development of chatbots.

NLP stands for **Natural Language Processing**, which is a branch of artificial intelligence that deals with the interaction between computers and human languages. NLP enables chatbots to understand and generate natural language texts based on user inputs or prompts. NLP also provides tools for text pre-processing, syntax analysis, and other advanced techniques that help chatbots handle human language.

GPT stands for **Generative Pre-trained Transformer**, which is a deep learning model that can generate natural language texts based on a given input or prompt. GPT is a type of language model, which is a probability distribution model that allows a chatbot to predict the next word or series of words in a sentence. GPT is trained on a large corpus of text data from the internet, enabling it to learn from the nuances and diversity of human language.

Some of the roles that NLP and GPT play in the development of chatbots are:

- **NLP** enables chatbots to process and understand natural language inputs from users and generate natural language outputs that are coherent, fluent, and engaging.
- **GPT** enables chatbots to learn from a large corpus of text data from the internet and generate texts that are relevant, diverse, and novel.

Figure 7-3 Two competing chatbots: Google Bard and Microsoft Bing

What ChatGPT can do

To give you a better idea of what a chatbot like ChatGPT can do, the next five figures present some examples. To get the most from this chapter, though, you should get your own ChatGPT account and enter the same prompts that are in the figures. You should also try those prompts on Google Bard and Microsoft Bing to see how the results vary.

To get a ChatGPT account, go to the OpenAI web page for opening a ChatGPT account (chat.openai.com/auth/login), and sign up for a *free* account. To get started with Bard, use the Chrome browser to go to bard.google.com. And to get started with Bing, start the Edge browser and use the Bing search engine in chat mode.

As you've already seen, all three chatbots produce similar, but not identical, results. Because ChatGPT was first on the market, though, the examples that follow use ChatGPT.

Chat

The idea of a "chat" is that there is continuity between a series of prompts and responses. This is illustrated by the example in figure 7-4. Here, the prompts are in boldface, and the responses are in regular type, as they will be in the rest of this chapter.

In this example, the first prompt asks a question about the use of iPads. Then, the next two prompts ask follow-up questions, and ChatGPT realizes that they're related to the previous prompts, as in a conversation.

In this case, all three responses start with an introductory paragraph that is followed by a numbered list, a bulleted list, or both. This is indicated by the parenthetical at the end of each response. In other words, the responses are much longer than what's shown here. To see the complete responses, enter the prompts in this figure in the chatbot that you're using. But keep in mind that the results won't be exactly the same, even if you're using ChatGPT.

I use an iPad about 4 hours each day. Will that cause any health problems?

Using an iPad for four hours each day is within a moderate range of screen time, and it may not necessarily cause immediate or severe health problems. However, it's important to be aware of potential risks and take precautions to maintain your well-being. Here are a few considerations:

1. Eye strain: Staring at a screen for prolonged periods can cause eye strain, dryness, and fatigue. To mitigate this, follow the 20-20-20 rule: take a 20-second break every 20 minutes to look at something 20 feet away. Additionally, ensure proper lighting and adjust screen brightness and contrast settings for comfortable viewing.
2. Sleep disturbances: Exposure to the blue light emitted by electronic devices, including iPads, close to bedtime can disrupt sleep patterns. It's recommended to limit iPad usage in the hours leading up to sleep or enable night mode features that reduce blue light emission.

(3 more numbered items, a bulleted list of 5 items, and a concluding paragraph)

Any special concerns for children 10 and under?

Yes, there are a few special concerns to consider regarding iPad usage for children aged 10 and under:

1. Developmental impact: Excessive screen time, including iPad usage, can potentially impact a child's development. At this age, children are in critical stages of physical, cognitive, and social development. Excessive screen time may interfere with activities that are important for their growth, such as active play, social interactions, and hands-on learning experiences.
2. Eye health: Children's eyes are still developing, and excessive screen time, especially at close distances, may contribute to eye strain, dryness, and potential vision problems like myopia. It's important to encourage breaks, outdoor activities, and a balanced use of digital devices to promote healthy eye development.

(4 more numbered items and two concluding paragraphs)

Any thoughts on social media time?

When it comes to social media time, it's important to approach it with caution and establish healthy habits, especially for children and teenagers. Here are some thoughts to consider:

1. Age appropriateness: Different social media platforms have age restrictions in place, and it's important to adhere to these guidelines. Many social media platforms require users to be at least 13 years old. For younger children, it's generally recommended to wait until they are mature enough to understand the potential risks and responsibly navigate social media platforms.
2. Parental guidance and supervision: For children and teenagers who are allowed to use social media, parental guidance and supervision are crucial. Parents should have open

(3 more numbered items and a concluding paragraph)

Figure 7-4 What ChatGPT can do: Chat

Parlor tricks

Figure 7-5 gives a few more examples that show what ChatGPT can do. But these are more frivolous. In fact, one analyst described uses like the ones in the first two examples as "parlor tricks."

Here, the first example writes a love poem in the style of Andrew Marvel who I chose because he wrote a poem called "To his coy mistress." Then, the second example answers a question about a penny dropping from the Empire State Building. Both of these are impressive, even if they are parlor tricks.

The third example shows that ChatGPT can translate English into German. That's also impressive, and that's more than a parlor trick.

The fourth example shows that ChatGPT will refuse to respond to any prompt that doesn't lead to a response that's safe and respectful. In this case, the prompt is harmless, but the guidelines are good.

Write a poem in the style of Andrew Marvell that asks my friend Nancy for a date with some flirty overtones.

Fair Nancy, let me take your hand,
And lead you to a night so grand,
With stars that shine, and breeze that blows,
And whispers soft, of sweet repose.

We'll wander through the fields, so green,
And bask in nature's beauty serene,
Our hearts afire, with love's sweet flame,
And revel in the joy that cannot be tamed.

Our laughter will fill the air so light,
And every step, will lead us to delight,
With each new moment, more divine,
Our love will only continue to shine.

So come, my love, and be with me this night,
Together we'll explore love's sweet delight,
And find in each other, what our hearts so desire,
For in this moment, thou art all I admire.

If a penny is dropped from the top of the Empire State Building and it hits a car below, what will happen to the car?

If a penny was dropped from the top of the Empire State Building and hit a car below, the penny would most likely cause minimal damage to the car. Despite its height, a penny's small size and low mass make it unlikely to cause significant damage to a car's body. At most, the penny might cause a small dent or scratch, but it would not significantly harm the car's structural integrity.

How do you say where is the railroad station on Freidrich Street in German?

In German, you would ask "Wo ist der Bahnhof an der Friedrichstraße?" to ask where the railway station is on Friedrich Street.

Write a poem in the style of Andrew Marvell that asks my friend Nancy for a date with some lustful overtones.

I'm sorry, but I cannot comply with this request. Creating poetry with lustful overtones is inappropriate and goes against OpenAI's guidelines for creating safe and respectful content.

Description

- The first two examples show the broad range of prompts that ChatGPT can handle.
- The third example shows how ChatGPT can translate from one language to another.
- The last example shows that ChatGPT won't cross certain boundaries.

Figure 7-5 What ChatGPT can do: Parlor tricks

Write programming code

This is beyond the scope of this writing book, but figure 7-6 shows how you can use ChatGPT to write programming code. This just demonstrates another of the many capabilities that ChatGPT has. In both examples, the text is also generated by ChatGPT.

Here, the first example shows the Python code that ChatGPT delivers when it is asked to write code that calculates the future value of $24 compounded at 3 percent interest from 1624 to the present. If you know anything about programming, the result is an interesting way to code this because it uses an exponent (the pow method) to calculate the answer when a more obvious way would be to loop through the years.

The second example shows the Java code for listing the prime numbers between 0 and 100. This time, the code uses a for loop within a for loop to determine whether a number is prime. And like the first example, this is impressive.

In fact, AI programs, including chatbots, are already changing the way programs are developed. So instead of writing code, the programmer's focus is on planning, coordinating, and testing the blocks of code that are written by AI. Of course, you still need to understand the code, but the increase in productivity has already led to the loss of thousands of jobs.

Write Python code to calculate the future value of $24 with 3% yearly interest compounded from the year 1624 until the present.

Here is an example of Python code that calculates the future value of $24 with 3% yearly interest compounded from the year 1624 until the present:

```
import datetime
import math

start_year = 1624
end_year = datetime.datetime.now().year
principal = 24
rate = 0.03

years = end_year - start_year
future_value = principal * math.pow((1 + rate), years)

print(future_value)
```

Keep in mind that this is a simplified example that doesn't take into account inflation or other economic factors that would affect the value of money over time.

Write Java code that will list the prime numbers between 0 and 100.

```
public class PrimeNumbers {
    public static void main(String[] args) {
        // List to store prime numbers
        ArrayList<Integer> primes = new ArrayList<Integer>();

        // Iterate through numbers between 0 and 100
        for (int i = 2; i <= 100; i++) {
            boolean isPrime = true;
            for (int j = 2; j <= i/2; j++) {
                if (i % j == 0) {
                    isPrime = false;
                    break;
                }
            }
            // If number is prime, add to list
            if (isPrime) {
                primes.add(i);
            }
        }
        // Print list of prime numbers
        System.out.println(primes);
    }
}
```

This code creates an ArrayList called `primes` where the prime numbers will be stored. Then, it iterates through numbers between 2 and 100. For each number, it checks if it's prime by dividing it by numbers from 2 to its half. If the remainder is 0, it means the number is not prime, so the flag `isPrime` is set to false. If the number is prime, it gets added to the `primes` list. Finally, it prints the list of prime numbers.

Figure 7-6 What ChatGPT can do: Write programming code

Research

In figures 7-2, 7-3, and 7-4, you've seen how a chatbot can be used for research. Now, figure 7-7 presents two more examples.

In the first example, the prompt asks ChapGPT how to set the salary for a programming job. That information can be used as a starting point for deciding how best to reorganize a company's hiring procedures. In the second example, the prompt asks ChatGPT a question about AI writing.

Of course, you can get the same information by using a search engine and going to the links it provides. But using a chatbot is an easy way to get research in a single response...even though you may have to do other research to verify that information.

How do you set the salary for a programming job?

The salary for a programming job can be set through a combination of several factors, including:

1. Market demand: The demand for programming skills in a particular area and the supply of qualified candidates will affect the salary range for programming jobs.
2. Experience and skills: A programmer's level of experience, education, and specific technical skills can influence their salary.
3. Company size and budget: The size and financial stability of the company, as well as their budget for salaries, can also impact the salary offered.
4. Location: Geographic location can also play a role in determining salary, with salaries in larger cities and technology hubs generally being higher.
5. Industry: The industry in which the programmer is working can also affect their salary, with certain industries, such as finance or healthcare, offering higher salaries for programming jobs.

Ultimately, the salary for a programming job is negotiable and can be influenced by many different factors. It is important for both the employer and employee to have an open and honest conversation about salary expectations to reach a mutually acceptable agreement.

How long have AI writing programs been in use?

AI writing programs have been in use for several decades, but the technology has become much more advanced and widely used in recent years. The earliest AI writing systems were developed in the 1950s and 1960s, but they were limited in their capabilities and not widely used.

In the 1990s and 2000s, advances in AI and NLP led to the development of more sophisticated AI writing systems that could handle more complex writing tasks. These systems were used primarily in specialized domains, such as financial reporting and technical writing.

In the past decade, the growth of big data and the development of deep learning algorithms have enabled the creation of much more powerful AI writing systems. These systems can generate high-quality text for a wide range of applications, including news summaries, social media updates, and customer support.

As AI and NLP technologies continue to evolve, it is likely that AI writing systems will become even more widely used and capable in the future.

Figure 7-7 What ChatGPT can do: Research

Write an essay or short report

Figure 7-8 shows a report written by ChatGPT when asked to write a report called "Why some eLearning courses fail and how to make sure your courses succeed." Now, imagine writing that report just by writing the prompt and waiting for several seconds. Wow! At first glance, that is impressive.

But take a closer look. Of course, some of the reports that ChatGPT generates are better than others. And some are okay. But this isn't ChatGPT's best work.

Here, the introduction is a general, non-committal paragraph, and the conclusion is a paragraph that summarizes the paragraphs between the introduction and the conclusion. So, to improve this report, you would start by improving the introduction and the conclusion.

What's worse, though, are the middle paragraphs. Here, four reasons are given for why eLearning fails and four tips are given for how eLearning can succeed. The trouble is that the tips just restate the reasons for failure. That's an unnecessary redundancy that wastes the reader's time and doesn't advance the argument.

Beyond that, the four points in both lists don't stand up to scrutiny. Does eLearning fail because of lack of engagement? I don't think so because eLearning can be quite engaging. Fun, even. Does eLearning fail because of poor instructional design? It can, but that's true of any course. That isn't the fault of eLearning. The same with the inadequate feedback and technological issues. They aren't the fault of eLearning. And what this report doesn't even address are the limitations of the primary delivery systems for eLearning: videos and slideshows.

So, yes, ChatGPT can write an essay or short report. But most of the time, you will have to abridge or enhance the content. And you will have to edit the response for clarity.

Write a report called: Why some eLearning courses fail and how to make sure your courses succeed

Introduction

In recent years, eLearning has become a popular way for individuals and organizations to learn and develop new skills. However, not all eLearning courses are successful. Some courses fail to achieve their objectives, resulting in low engagement and poor learning outcomes. In this report, we will explore the reasons why some eLearning courses fail and provide tips on how to make sure your courses succeed.

Reasons for Failure

1. Lack of engagement: One of the main reasons why eLearning courses fail is a lack of engagement. Learners may lose interest in the course due to its unappealing design or lack of interactive features.
2. Poor instructional design: Poor instructional design can make it difficult for learners to understand the content. The course may not be well-organized, or the information may be presented in a confusing or overwhelming manner.
3. Inadequate feedback: Learners need feedback to know if they are on the right track. If the course does not provide feedback or provides inadequate feedback, learners may become frustrated and disengage from the course.
4. Technological issues: Technical issues such as slow loading times or malfunctioning features can discourage learners and hinder their learning experience.

Tips for Success

1. Engaging design: A visually appealing and interactive design can help keep learners engaged. The course should include multimedia elements such as videos, images, and animations.
2. Clear instructional design: Clear instructional design can make the course easy to navigate and understand. The content should be organized into small, manageable chunks and presented in a logical and sequential manner.
3. Effective feedback: Feedback should be provided throughout the course, highlighting both correct and incorrect answers. Feedback should be timely and specific, helping learners understand where they went wrong and how to improve.
4. Technological stability: Technical issues can be avoided by thoroughly testing the course before launch. The course should be optimized for different devices and internet speeds, and the platform used for hosting the course should be reliable.

Conclusion

The success of an eLearning course depends on various factors such as design, instructional design, feedback, and technological stability. By implementing these tips, course creators can ensure that their courses are engaging, well-designed, and provide effective learning outcomes. Ultimately, the goal is to create courses that not only meet the needs of learners but also achieve their objectives.

Figure 7-8 What ChatGPT can do: Write an essay or short report

What ChatGPT can't do

Now that you know what ChatGPT can do, let's look at what ChatGPT can't do. Or to say that another way, let's see where ChatGPT fails.

Write like a pro

In the last figure, you saw an example of a ChatGPT report that isn't written at a professional level. Now, figure 7-9 presents another one. This time, the prompt asks ChatGPT to compare eLearning with traditional classroom instruction, especially when teaching how to program in Python.

Here, the first and last paragraphs in the response say nothing of value. The generalizations in the middle paragraphs aren't supported by specifics, and they don't do an adequate job of explaining the differences between eLearning and traditional classroom instruction. Besides that, none of the paragraphs pertain to the special requirements of teaching Python programming, even though our research shows that eLearning doesn't work for skills like that. To compound those problems, the Reading Grade Level for the report is 15.6, which further obscures the response.

In fact, if you look at enough examples, you have to conclude that ChatGPT can't write like a pro. Instead, the responses usually start with a paragraph that restates the prompt, followed by three or more numbered or bulleted items, followed by a concluding paragraph that restates the items in the numbered or bulleted list. That may be okay for a high school essay, but there's a lot more to professional writing than that.

Compare eLearning with traditional classroom instruction when you're teaching how to program in Python.

When teaching how to program in Python, both eLearning and traditional classroom instruction have their strengths and weaknesses. Each approach has its own unique benefits that make it better suited for certain learning styles and goals.

One of the main advantages of eLearning for programming in Python is that it allows students to work at their own pace. This is particularly helpful for students who have prior programming experience or for those who need to move more quickly through the basics in order to get to the more advanced topics. eLearning also provides students with the ability to revisit concepts and examples at any time, which can be invaluable for those who need extra practice or who struggle to retain information.

In contrast, traditional classroom instruction provides students with the opportunity to receive immediate feedback from a teacher and to collaborate with classmates on programming projects. This type of interaction can be particularly helpful for students who need additional guidance and support, especially when dealing with complex programming concepts. In addition, traditional classroom instruction often provides a more structured learning environment that can help students stay on track and avoid getting overwhelmed by the material.

Another advantage of eLearning for programming in Python is that it offers students a wide variety of resources, including video tutorials, interactive simulations, and programming exercises. This can be particularly helpful for students who learn better through hands-on experience and who need additional practice with coding exercises. eLearning also provides students with the ability to access educational content from anywhere and at any time, which can be ideal for individuals who have busy schedules or who need to balance education with work or other responsibilities.

On the other hand, one of the main disadvantages of eLearning for programming in Python is that students may miss out on the personal interaction and direct feedback that they would receive in a traditional classroom setting. This can be especially challenging for students who are new to programming and who need additional support in order to grasp the material. Additionally, students who learn better through hands-on experience may struggle with the lack of interaction and collaboration that eLearning provides.

In conclusion, both eLearning and traditional classroom instruction have their advantages and disadvantages when teaching how to program in Python. While eLearning provides students with flexibility and a wealth of resources, traditional classroom instruction offers students the opportunity to receive immediate feedback and collaborate with classmates. Ultimately, the best approach will depend on the individual student's learning style, goals, and personal circumstances.

A few of the problems with this ChatGPT report

- It's repetitive, and doesn't take a position.
- It doesn't support its generalizations with specifics.
- Its Reading Grade Level (RGL) is 15.6.

Figure 7-9 What ChatGPT can't do: Write like a pro

Add new perspectives and analysis

Figure 7-10 presents another example that exposes ChatGPT. This time, the prompt is to specify the advantages that books have when compared to eLearning.

In the response, the first paragraph rephrases the prompt, and the last paragraph points out the advantages of eLearning. That's okay, but whether you use books or eLearning doesn't "depend on the learning style of the individual and the type of material that they want to learn." It's more complicated than that.

Now look at the items in the bulleted list. Aren't the first and fourth points the same: Offline access and Low-tech? And the second point: Tangibility? Wow. Cost effective? Debatable. Reading experience and Lasting value? Okay. But overall, that's a mighty poor list. Silly even.

What's worse is that this response misses the important advantages that books have over eLearning. For instance, research suggests that most people read faster and retain more when they read books. It's obvious that eLearning provides for only a small fraction of the knowledge and skills that you need in life. And my own testing shows that the more difficult the skill is, the less effective eLearning is likely to be.

As I see it, this is the biggest weakness of ChatGPT. Apparently, it can't add perspectives and analysis that isn't somewhere in the thousands of documents that it has been trained on. That's why original thinkers will always be valued over chatbots.

What advantages do books have when compared to eLearning?

Books have a few advantages when compared to eLearning:

1. Offline access: Books do not require an internet connection to be read, whereas eLearning typically requires an internet connection to access the material.
2. Tangibility: Books have a physical form which can be held, annotated and passed on, which can be a more personal and satisfying experience for some people.
3. Reading experience: Reading a book can be a more immersive and enjoyable experience for some people, as it allows them to fully focus on the material and engage with it at their own pace.
4. Low-tech: Books don't require any special equipment or software, they can be read with a simple light source and don't require any power source.
5. Cost-effective: Books are typically less expensive than eLearning, particularly when considering the cost of access to eLearning material over time.
6. Lasting value: Books can be kept as reference materials for future use, and can also be re-sold.

That being said, eLearning also has many advantages such as interactive features, multimedia, instant feedback, and the ability to track progress, so it depends on the learning style of the individual and the type of the material that they want to learn.

A few of the points that ChatGPT is missing

- Most people can read 3 to 5 times faster than they can watch a video or go through a slide show.
- Most people retain more when they read than they do when they're watching a video or going through a slide show.
- eLearning provides for only a small fraction of the information and skills that are available through books.
- Books usually work better for teaching complex subjects.

Description

- Remember that ChatGPT is trained by thousands of documents so it can't add perspectives or analysis that isn't there.

Figure 7-10 What ChatGPT can't do: Add new perspectives and analysis

Write comprehensive reports

The last way that ChatGPT fails is that it can't write comprehensive reports that consist of more than a couple topics. That's especially true if the report requires subtopics.

As you learned in chapters 1 and 2, topics should be identified by headings and subtopics by subheadings...and ChatGPT isn't designed to do either. Yes, the report in figure 7-8 shows that ChatGPT has a rudimentary ability to include headings, but that's inadequate for professional writing.

To illustrate this problem, the first example in figure 7-11 shows the headings and subheadings for a report on eLearning. It consists of 5 headings with a few subheadings after each of the first four headings. As you can see, that's a plan for a comprehensive report on eLearning.

Then, the second example presents the headings and subheadings for this chapter. Right now, for example, you're reading the text for the third subheading under the heading, "What ChatGPT can't do."

Can you image ChatGPT developing the heading plans for reports or chapters like these? I can't either. Especially because the text in the training data isn't likely to use headings and subheadings in an effective way. Why? Because that's one of the shortcomings of most business writing.

The heading plan for an eLearning report

- **Introduction to eLearning**
  - **What eLearning is**
  - **The primary components of eLearning**
  - **eLearning and the Learning Management System (LMS)**
- **A quick tour of some typical eLearning courses**
  - **MOOCs**
  - **Commercial video courses**
  - **Corporate eLearning courses**
  - **College eLearning courses**
- **Why most eLearning courses fail**
  - **The limitations of video**
  - **The limitations of slides**
  - **The missing ingredients**
- **Why our eLearning courses will succeed**
  - **Our proven books will be the primary delivery system**
  - **Our eLearning components will add to the effectiveness of our books**
- **My recommendation**

The heading plan for this chapter

- **Introduction to AI writing and ChatGPT**
  - **An introduction to AI writing**
  - **How ChatGPT has taken AI writing to a new level**
  - **Two competing chatbots: Google Bard and Microsoft Bing**
- **What ChatGPT can do**
  - **Chat**
  - **Parlor tricks**
  - **Write programming code**
  - **Research**
  - **Write an essay or short report**
- **What ChatGPT can't do**
  - **Write like a pro**
  - **Add new perspectives and analysis**
  - **Write comprehensive reports**
- **How to use ChatGPT to improve your writing**
  - **Write prompts that get the intended results**
  - **Use ChatGPT for research...but do your own research too**
  - **Use ChatGPT for drafts...but enhance, edit, or ignore them**
- **Perspective**

Description

- At least for the time being, ChatGPT can't write a comprehensive report that requires more than a few topics (headings). And that's especially true for reports that require both topics and subtopics (headings and subheadings).

Figure 7-11 What ChatGPT can't do: Write comprehensive reports

How to use ChatGPT to improve your writing

Now that you have some idea of what ChatGPT can and cannot do, you should try it. That will help you decide whether it can help you write faster and better. To do that, you can not only use ChatGPT to do research, but also to write first drafts. To get the most from either of those capabilities, though, you need to write prompts in a way that gets the intended results. So let's start with that.

Write prompts that get the intended results

It may not be obvious so far, but how you write the prompt is critical if you want to get the intended results. In fact, that's so important that some companies have employees who are trained to be *prompt managers*. Their job is to write the prompts that get the results that are needed. But how do you write those prompts?

Just for fun, I asked the three chatbots what their guidelines are for writing effective prompts. The Bing response is shown in figure 7-12 because I thought it was the best one, even though "Be creative" makes no sense at all. Otherwise, who can argue with advice like: Start with a clear goal. Be brief and to the point. Avoid unnecessary words. Be conversational. Avoid jargon and slang. Be specific.

Because that response is unnecessarily wordy, it is followed by my response. First, remember that if your prompts are easy for humans to read and understand, they will also work with chatbots. Second, if you read chapter 4 in this book, you will learn all the skills that you need for writing prompts that are easy to read and understand. Third, if you write a prompt that doesn't get the result you want, you should analyze the prompt, improve it, and try it again.

Incidentally, prompts can be as long as they need to be to get a useful result. Here, for example, is a prompt that I used to see whether I could get one of the chatbots to produce a useful report on eLearning:

> Write a report that answers the question: Should we use eLearning in our training programs? The report should explain what eLearning is, describe the eLearning components, and explain how eLearning and the LMS work together. It should also describe the types of subjects for which eLearning works the best as well as the types of subjects for which eLearning usually fails. If possible, it should list specific subjects for which eLearning works and doesn't work.

If you try this prompt on ChapGPT, Bard, and Bing, you'll see that all three responses provide a lot of information, but all have the same shortcomings that the other chatbot responses in this chapter have: too much generalization without the specifics that bring the ideas to life...and no useful analysis or new perspectives. That's why it's usually better, as the Bing response in this figure says, to keep your prompts concise.

What are the best guidelines for writing chatbot prompts?

The Bing response

Chatbot prompts are the messages that initiate or guide a conversation with a chatbot. They can be questions, statements, commands, or suggestions that elicit a response from the chatbot. Writing effective chatbot prompts can improve the user experience and the chatbot performance.

Some of the best guidelines for writing chatbots are:

Start with a clear goal: Determine what you want the chatbot to achieve through the prompt. This could be providing information, guiding the conversation, or solving a problem. If your question is vague, your answer will be vague.

Keep it concise: Keep the prompt brief and to the point. Avoid unnecessary words or details that could confuse the chatbot or the user. Use simple and direct language that is easy to understand and follow.

Use natural language: Write the prompt in a way that is conversational and friendly. Use everyday language that matches the tone and style of your chatbot and your target audience. Avoid jargon, slang, or technical terms that could be unfamiliar or unclear to the chatbot or the user.

Avoid ambiguity: Ensure that the prompt is specific and avoids ambiguity. Use keywords or phrases that are relevant to the topic and the context of the conversation. Provide clear instructions or expectations for the chatbot or the user. Use examples or clarifications if needed.

Be creative: Use prompts that are engaging and interesting to the user. Use humor, emotion, or personalization to make the conversation more lively and human-like. Use prompts that encourage exploration, discovery, or feedback from the user. Use prompts that challenge or inspire the user to think creatively or critically.

My response

- If your prompts are easy for humans to read and understand, they will get the best chatbot results.
- To write prompts like that, use the techniques that are presented in chapter 4, "How to write sentences that are easy to read and understand."
- When you write a prompt that doesn't get the intended result, improve it and try again.

Description

- Your prompts can be as long as they need to be to describe what you want ChatGPT to produce. However, the longer the prompt is, the harder it is to keep it specific enough to get the results that you want.

Figure 7-12 Write prompts that get the intended results

Use ChatGPT for research... but do your own research too

The more you use a chatbot, the more you realize what a powerful tool it can be for research. In fact, I used ChatGPT to do some of the research for this chapter. And some of my prompts are listed in the first group in figure 7-13.

However, as you've already seen, ChatGPT's responses aren't always complete...and they aren't always accurate either. In fact, some responses are so far from accurate that they're called *hallucinations*. That's why you always need to do your own research too. You can't just rely on what you get from a chatbot.

To illustrate, the second set of bulleted items in this figure lists a few examples of other research that I did for this chapter. That includes using a search engine to do more research and check facts. It includes experimenting with other AI writing products to see how they work and what they can do. And it includes experimenting with different types of writing (or use cases). In fact, it took far more time to do the research for this chapter than it took to write it. But until you do research like that, you really don't know enough to make reasonable judgements.

The third and fourth lists in this figure are examples of the research you might do for a report on eLearning. It usually makes sense to start by using ChatGPT to get responses to prompts like those in the third list. But here again, you've already seen that ChatGPT's responses are likely to be incomplete.

That's why you need to do your own research along the lines of what's in the fourth list. First, use a search engine like Google (not Google Bard) to do more research and fact checking. But don't stop there. Take a few eLearning courses to see how well video and slideshows work for instruction. Take an eLearning course on a subject you already know, and take eLearning courses on both simple and complicated subjects. You might even want to buy a book on one of the subjects to see how well it works when compared to an eLearning course that you've taken. Only then are you prepared to draw your own conclusions.

Typical ChatGPT prompts for research on AI writing

- What are the names of some of the best AI writing products?
- What are some common uses of AI writing?
- What are the technologies that drive ChatGPT?
- How long have AI writing programs been in use?

Other research that I did for this chapter

- Use a search engine to check facts and do more research.
- Experiment with more than one AI writing product.
- Experiment with several different types of writing.
- Evaluate the responses.

Typical ChatGPT prompts for research on eLearning

- What are the primary components of an eLearning course?
- How does taking an eLearning course compare with reading a book on the subject?
- What are the factors that determine the success of an eLearning course?
- How does video compare with a slideshow when it comes to programmer training?
- Compare eLearning with traditional classroom instruction.

What you might do as part your own research on eLearning

- Use a search engine to check facts and do more research.
- Take an eLearning course that's based on video to see how that works.
- Take an eLearning course that is based on a slideshow to see how that works.
- Take an eLearning course on a subject you already know.
- Take an eLearning course for a simple skillset as well as a more complicated skillset.
- Compare what you learn in an eLearning course with what you learn when you read a book.

Description

- For some subjects, ChatGPT provides an efficient way to do some research.
- But you also need to do your own research to make sure that the ChatGPT responses are complete and accurate.

Figure 7-13 Use ChatGPT for research...but do your own research too

Use ChatGPT for drafts... but enhance, edit, or ignore them

Okay, you've done your research. Should you now use ChatGPT for writing the first draft? Figure 7-14 presents some ideas that can help you decide.

For short reports, you may as well give ChatGPT a try to see what it can do. Since you've already done your research, you'll be able to decide whether the response is good enough to enhance or edit.

For longer documents, you should start by planning the headings and subheadings as shown in chapters 1 and 2. Then, you can issue prompts for some of the topics or subtopics within the document to see whether they might be useful.

For most documents, though, you will need to enhance the first draft or drafts by adding what ChatGPT has missed. Then, when you're sure that all the required ideas are in the draft, you will need to edit the draft. And that editing will probably be extensive if you want to raise the writing to a professional level.

The trouble is that editing someone else's writing can be more difficult than writing the document in the first place. That's true even when the writer that you're editing is ChatGPT. So if this process gets too frustrating, you may decide to ignore what ChatGPT gives you and write the document from scratch.

In fact, I used ChatGPT to do some of the research for this chapter, but I didn't get any responses that I could use as portions of a first draft. I've also asked several of my business and technical friends who have been monitoring the development of ChatGPT if they have used or would use ChatGPT for their own writing. So far, no one has said, yes. And that's especially true for those who write long documents that require headings and subheadings.

My suggestion, then, is to use ChatGPT for research, but do your own research too. Then, see whether ChatGPT can deliver a good first draft. If not, just write the document from scratch. That will not only be more satisfying, but you'll end up with a better document.

Use ChatGPT to write the first draft or drafts

- For short reports and essays, you may be able to issue a single ChatGPT prompt and get a response that will work as a first draft.
- For longer documents, you should plan the headings and subheadings as shown in chapters 1and 2. Then, you can issue prompts that address the different topics or subtopics.

If necessary, abridge or enhance the first draft or drafts

- If a first draft doesn't present the ideas that your research shows are needed, you need to abridge or enhance the first draft.
- If you've issued multiple prompts for information, you need to combine the responses into a cohesive document.

Edit the first draft

- Edit the first draft so it's consistent with the writing standards that are presented in this book.

If this gets too frustrating, write the document from scratch

- Working with someone else's writing can quickly become irritating and counter-productive. In that case, it's often easier to use the methods in this book to write the document from scratch.

Description

- For simple prompts, you may be able to edit what ChatGPT writes and be done with it.
- But for most prompts, you will need to abridge, enhance, and edit the responses. If that gets too cumbersome, you may decide that it's easier to start from scratch.

Figure 7-14 Use ChatGPT for drafts...but enhance, edit, or ignore them

Perspective

Chatbots like ChatGPT are a major technological breakthrough that's in the tradition of the internet, the iPhone, and social media. That's why chatbots are changing the way people are doing their research and writing. And these products, including Google's Bard and Windows Bing, are only going to get better.

That's why it's worth taking the time to experiment with chatbots. See how your prompts affect the results that are produced. See whether chatbots can speed up your research. See whether you can get chatbots to produce first drafts that you can edit and be done. Only then can you decide whether chatbots can help you write faster or better.

Just know that you can't just trust that your chatbot will present all the information, the right information, or even accurate information. Also know that you will usually have to edit or rewrite what a chatbot produces because it won't be written at a professional level. In short, if you want to write like a pro, you still need to master the tested writing methods that are presented in this book.

Terms

AI (Artificial Intelligence)
AI writing
SEO (Search Engine Optimization)
chatbot
chatbot prompt
chatbot response
prompt manager
hallucination

Summary

- *Artificial Intelligence* (*AI*) has been used for *AI writing* for many years, but *chatbots* like ChatGPT, Google Bard, and Microsoft Bing are taking AI writing to a new level.
- Although chatbots have many capabilities, doing research and writing first drafts are the capabilities that matter most to writers.
- When you treat a chatbot like ChatGPT as a black box, you can analyze its limitations. That includes not being able to write at a professional level, add new perspectives and analysis, and write comprehensive reports.
- To get the most from chatbots, you need to be able to write *prompts* that get the intended *responses*. To do that, you need to use the methods in chapter 4 in this book to clearly state what you want the chatbot to produce.
- If you use a chatbot to do research, you need to make sure that the chatbot responses are complete and accurate.
- If you use a chatbot to write first drafts, you need to be able to abridge, enhance, and edit them.

Section 2

The Microsoft Word features for writers

If it seems strange to include three chapters on Microsoft Word in a book on writing, that's how important we think the Word features for writers are. Templates and styles will help you format your documents faster and better than ever. The outline feature works great for planning the headings and subheadings for a document. And the spelling and grammar checker will not only help you find and fix a few errors, but also produce a grade level score for each document so you can see whether you need to improve its readability.

So in this section, chapter 8 shows you how to use Word's templates and styles. Chapter 9 shows you how to use Word's outline feature. And chapter 10 shows you how to use Word's spelling and grammar checker. If you do much writing and you aren't already using Word, these chapters just may convince you to switch to Word right away.

As you read these chapters, you may find that there are a few differences between the version of Word that you're using and the one used for the examples in this section. If so, the differences should be so minor that you shouldn't have any trouble adjusting to them.

8

How to use Word's templates and styles

To my continued surprise, many business writers don't use templates and styles for documents like reports, proposals, and resumes. How do I know that? Because they send me documents that are poorly and inconsistently formatted. That's even true for people who are applying for writing jobs!

If you're one of these people, this chapter will show you how using templates and styles will not only improve the formatting in your documents, but also your writing productivity. So to start, this chapter shows you how to use existing templates and styles, which every writer should know how to do. Then, this chapter shows you how to create and modify templates and styles in case you also need to do that.

Basic concepts

A *template* is a document that is used for starting other documents. Among other items, a template includes the *styles* that are used for formatting the paragraphs in a document. This chapter starts by presenting the basic concepts for using templates and styles.

Template concepts

When you use Word, all documents are started from templates. But if you don't specify the template that you want to use, the Normal template is used. That template has no headers, footers, or starting text, but it does have a collection of *built-in styles* that are shown in the Styles group of the Home tab in the Word ribbon.

When you prepare a document like a report or proposal, though, you should start it from a template that includes the basic formatting for the document. That usually includes a header or a footer that includes the page number, the file name, and the date. The template should also include the *styles* for formatting the paragraphs in the document. And it may include *Word fields* that generate items like a *table of contents*.

To illustrate, figure 8-1 presents a document that has just been started from a template named "Report." In the Styles pane, you can see a list of the styles that this template provides. In the document window, you can see that this template includes starting text that shows the formatting and shortcut keys that the styles provide. Although it isn't shown, this template also has a footer that contains the document name, date, and page number.

To use this template, you replace "Report title" with the report title that you're going to use. You also delete all of the paragraphs that follow after you see what the style formatting looks like and what the shortcut keys are. Then, you can start writing, and you can apply the styles in the Styles pane whenever they're needed.

In short, with just a few keystrokes or mouse clicks, you can start a document that has the page formatting, starting text, headers, footers, and styles that you need. Then, you can apply the styles to the paragraphs that you write, and thus make sure that the formatting is consistent throughout the document.

A document started from a Report template

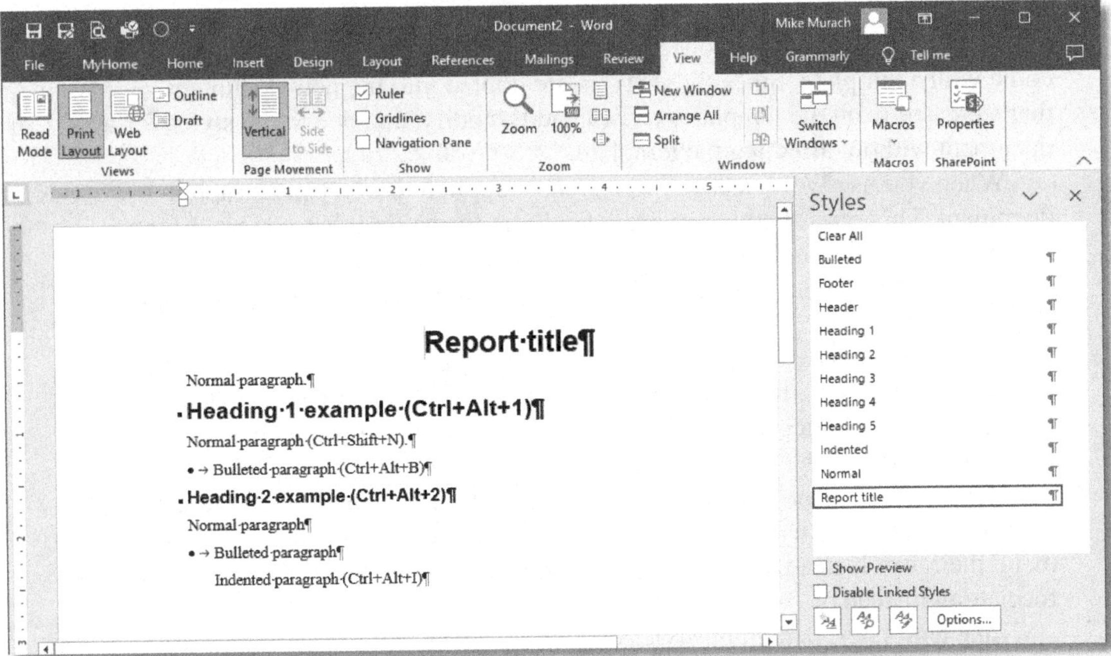

Template concepts

- A *template* is a document that is used for starting other documents. The template includes the *styles* that are used for formatting the paragraphs in the document.
- Besides styles, a template can include page formatting, headers and footers, starting text and graphics, *Word fields*, and more.
- All Word documents are started from a template. If you don't start a new document from a specific template, the document is based on the Normal template, but all that template offers is some *built-in styles* that aren't of much value.

The benefits that you get from using templates

- With just a few keystrokes or mouse clicks, you can start a document that has all the page formatting, starting text, headers, footers, and styles that you need.
- Templates enforce consistent formatting for each type of document.

Figure 8-1 Template concepts

Style concepts

Figure 8-2 presents the concepts that you need for using the *styles* that come with a template. Since these styles are copied into the new document that's created from the template, you can add, modify, and delete the styles for a document without affecting the template.

When you use Word, a *paragraph style* is applied to every paragraph in a document. This type of style can provide all the formatting that you need for a paragraph including indentation, the spacing before and after each paragraph, tab settings, borders, bullets and numbering, and the font that's used for the characters in the paragraph.

In contrast to a paragraph style, a *character style* provides the formatting for selected characters within a paragraph. For instance, you can create a character style that applies italics to the selected text. You can also create a character style that changes any of the characteristics of the selected text.

However, if you're just applying italics or boldface to the selected text, there isn't much benefit to using character styles. So in that case, we don't recommend their use. Instead, as you'll soon learn, we recommend that you use direct formatting instead of character styles.

Like templates, paragraph styles encourage consistent formatting from one document to another. They also encourage consistent formatting within a document. And they help you work more efficiently because you can apply a style with just a mouse click.

When you start using styles, you may find it useful to open a style area to the left of the document you're working on. To do that, you can use the procedure in this figure. This area shows the name of the style that's applied to each paragraph. Note, however, that the style area only shows in Draft view and Outline view, not in Page Layout view.

In this figure, the Heading 2 style is applied to the first heading; the Normal style is applied to the paragraph that follows; the Bulleted style is applied to the next four paragraphs, and the Normal style is applied to the two paragraphs after that. In general, the template for a document should provide all of the styles you need for the type of document that you're writing. That way, the formatting will be consistent throughout the document.

A document that uses the styles in the Report template

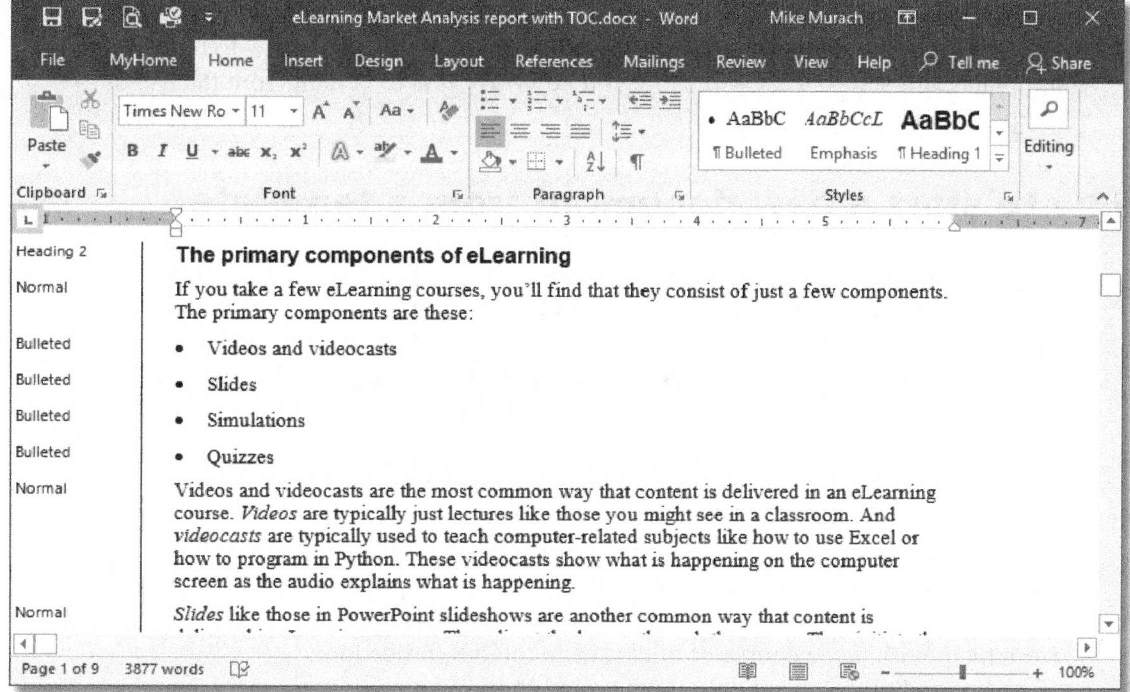

Style concepts

- A *paragraph style* controls both the paragraph and character formatting of the paragraph that the style is applied to.
- A *character style* controls just the formatting of the characters that it is applied to.
- When a new document starts, the styles from the template that it is based on are copied into the document. Then, you can add, modify, and delete styles in the document without affecting those in the template.

The benefits that you get from using styles

- With just a few keystrokes or mouse clicks, you can apply complex formatting to a paragraph. This also enforces consistent formatting.
- You can change the formatting for all the paragraphs that have the same style applied to them just by changing the formatting for that style.

How to display the style area that shows the style for each paragraph

- Use the File→Options command to open the Word options dialog box. Next, click Advanced to go to the advanced options, and scroll down to the Display group. Then, set the Style Area Pane Width in Draft and Outline View option to 1 inch.
- Use the View tab to switch to Draft view, and the style area will be displayed to the left of the document, as shown above. This area shows the name of the style that's applied to each paragraph.

Figure 8-2 Style concepts

How to use templates and styles

Now that you understand the concepts, this topic shows you how to use templates and styles. That will show you how easy it is to benefit from the use of templates and styles.

How to start a new document from a template

Figure 8-3 shows how to start a new document from a template. First, if the templates have already been loaded onto your computer, you can use the File→New command to access them. Although how you find the templates varies a bit from one version of Word to another, you should be able to find them without much trouble. Then, you can select the template you want to use and click on it.

However, before you can access the templates that way, you need to load them into the Custom Office Templates folder. To do that, you use the second procedure in this figure. In brief, you find the templates that you want with File Explorer, copy them, and paste them into the Custom Office Templates folder. Once your templates are loaded into that folder, you can start new documents from them with just a couple of clicks, as in the first procedure.

The third procedure shows another way to start a new document from a template. For that, you use File Explorer to find the template in a folder on your PC or server, and then double-click on it. To identify a template in File Explorer, you can look for the .dotx extension or the Word icon for templates. Of course, it's better to load the templates that you're going to use in the Custom Office Templates folder and access them with the New command, but this is okay for one-time or occasional use.

The Personal pane in the New window (File→New)

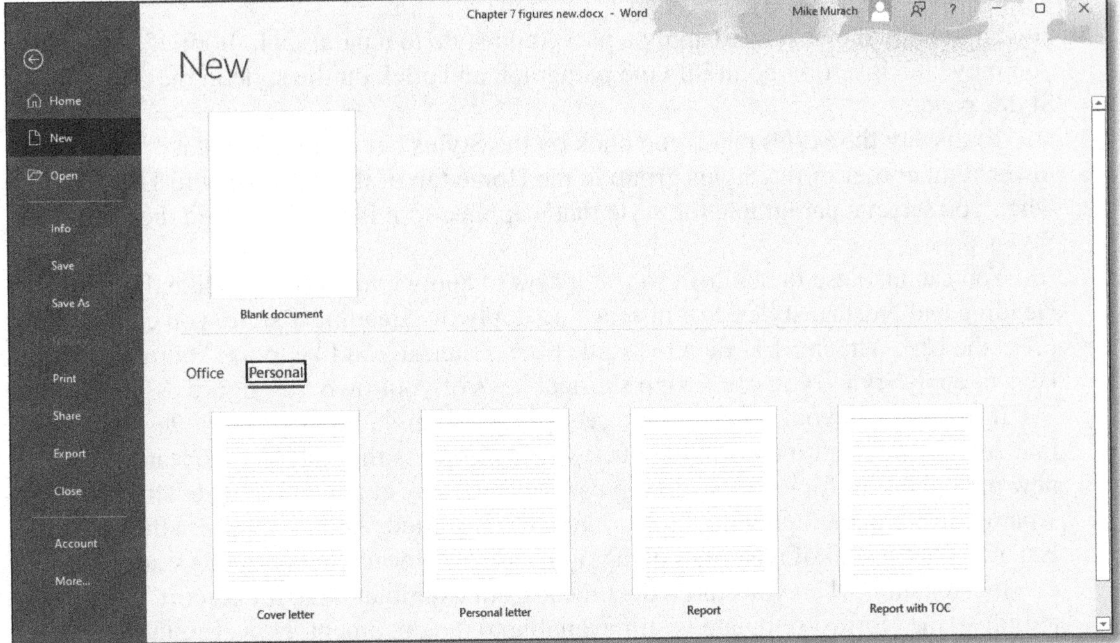

How to start a new document by using the File→New command

- Use the File→New command to show the available templates.
- If necessary, click on Personal to display your templates.
- Click on the template that you want to start the document from.

How to load templates into the Custom Office Templates folder

- Find the templates in File Explorer. Then, copy them to the clipboard.
- Find this folder in File Explorer and paste the templates into it:
  `\Documents\Custom Office Templates`

How to start a new document that's anywhere in File Explorer

- Find the template in File Explorer, and double-click on it. Word templates are identified by the .dotx extension.

Description

- If your company uses templates, you can load them onto your computer by using the second procedure above. Then, they will be available whenever you start a new document.
- The way your personal templates are displayed will vary slightly from one version of Word to another.

Figure 8-3 How to start a new document from a template

How to apply paragraph styles

Figure 8-4 shows how to apply a paragraph style to a paragraph. In brief, you move the insertion point into the paragraph and click on the style in the Styles pane.

To display the Styles pane, you click on the Styles button that is in the lower-right corner of the Styles group in the Home tab of the Word ribbon. Then, when you select a paragraph, the style that's applied to it is highlighted in the Styles pane.

You can also use the built-in *shortcut keys* to apply common styles like the Heading and Normal styles. For instance, to apply the Heading 1 style, you can press the Ctrl, Alt, and 1 keys at the same time. Later, if you like to use shortcut keys to apply styles, you can assign shortcut keys of your own (see figure 8-10).

If the styles in your templates are set up right, though, many of the styles in a document are automatically applied when you press the Enter key to start a new paragraph. If, for example, you press the Enter key at the end of a Heading 1 paragraph, a paragraph with the Normal style is started. And if you press the Enter key at the end of a Normal paragraph, another Normal paragraph is started.

Incidentally, when you start a document from a template like the Report template, the Normal template is still available to the document. However, the Report template takes precedence over the Normal template whenever both templates provide the same item. If, for example, you apply the Normal style to a paragraph in a document that's started from the Report template, the style is taken from that template, not the Normal template.

The Styles pane for a document that uses the Report template

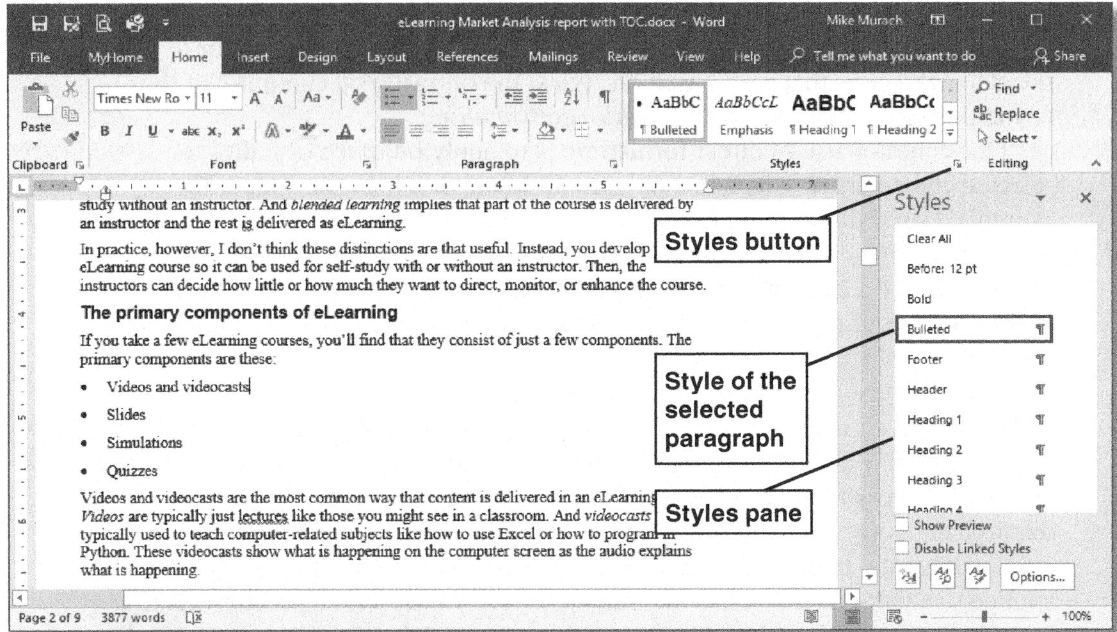

The primary styles in the Report template

Style name	Use	Shortcut keys	Next paragraph
Report title	Report title		Normal
Heading 1	Level-1 heading	Ctrl+Alt+1	Normal
Heading 2	Level-2 heading	Ctrl+Alt+2	Normal
Normal	Normal text paragraph	Ctrl+Shift+N	Normal
Bulleted	Bulleted items		Bulleted
Indented	Indented items		Indented

How to use the Styles pane to apply styles

- To display the Styles pane, click the Styles button in the Home tab.
- To apply a paragraph style, move the insertion point into the paragraph. Then, in the Styles pane, click the name of the style that you want to apply.

How to use the shortcut keys to apply styles

- By default, you can use the shortcut keys in the table above to apply the heading styles and the Normal style.
- You can also create your own shortcut keys for applying styles as shown in figure 8-10.

How styles are applied when you start a new paragraph

- Each paragraph style has a designated style for the paragraph that follows it when you press the Enter key at the end of the paragraph. However, if you press the Enter key in the middle of the paragraph, a paragraph with the same style is started.

Figure 8-4 How to apply paragraph styles

When and how to use direct formatting

Direct formatting means that you apply formatting to a paragraph or to selected characters that isn't based on styles. In contrast, you can think of the formatting that's done by styles as *indirect formatting*.

One common use of direct formatting is to apply boldface or italics to selected characters or words in a paragraph. In the previous paragraph, for example, I used direct formatting to italicize four of the words. That's usually better than using character styles, because it's as easy to apply boldface or italics as it is to apply a character style. Besides that, you don't have to create the character style. If the character formatting is more complex, however, character styles can be useful.

Occasionally, you may also need to use direct paragraph formatting. If, for example, you need to change the tab settings for a table style, you can use direct formatting for that. In general, though, the styles in your templates should provide for every type of paragraph formatting that's required. So if a new requirement comes up, you probably ought to create a new paragraph style for it.

To illustrate, figure 8-5 shows what happens when direct formatting is applied to a paragraph that has the Indented style. In this case, direct formatting has been used to indent the right side of the paragraph by .5 inches. If you look in the Styles pane, you can see that this caused a new style to be added to the pane (Indented + Right 0.5").

From that point on, you can click on that style if you want to format a paragraph that's indented from both the left and the right. If, however, you want to modify the original Indented style so the right indent is applied to all paragraphs with the Indented style, you can use the procedure in the next figure. Once that's done, the new style will be removed from the Styles pane. Often, that's the best way to handle a style change like this.

The styles pane after the indented paragraph has been modified

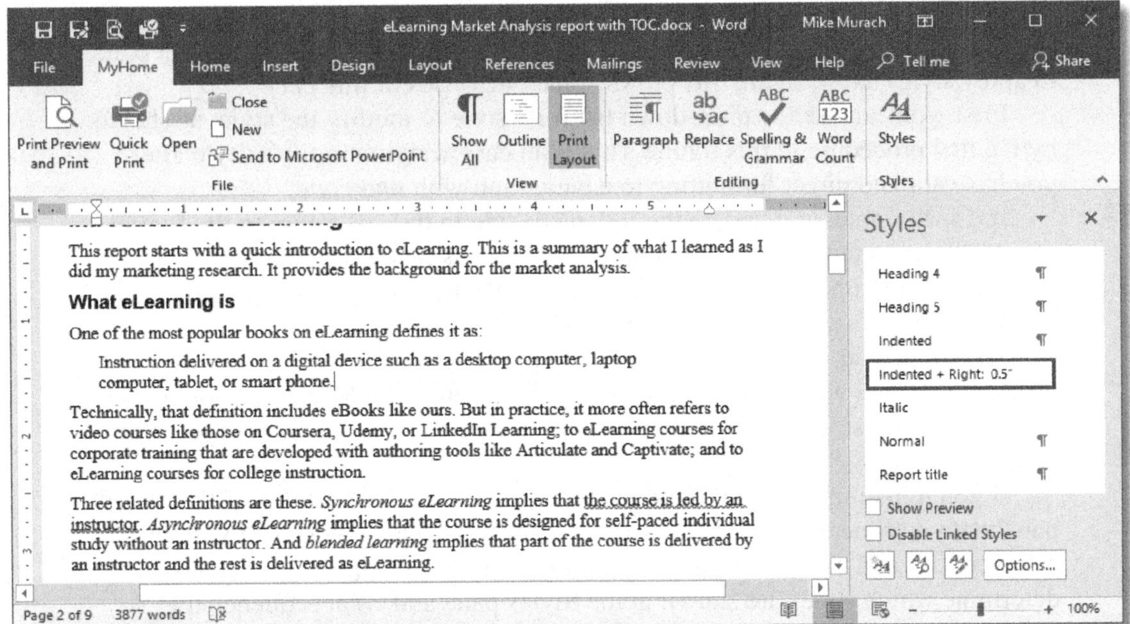

When to use direct formatting

- When you need to apply simple character formatting like italics within a paragraph.
- When you want to modify the formatting for just a paragraph or two and don't want to create a new style for that formatting.

What happens when you apply direct formatting to a paragraph

- Another style is added to the Styles pane that shows what modifications have been made to the original paragraph style. In the example above, a paragraph based on the Indented style has been indented from the right side by 0.5 inches so a style named Indented + Right 0.5" has been added to the Styles pane.

Two ways to handle the new style that has been added

- If necessary, you can apply the new style to other paragraphs.
- If you want all paragraphs of that type to have the changed formatting, you can change the original style so it's the same as the modified style, as shown in the next figure. Then, the style that had been added will be removed from the Styles pane.

Description

- If the template you're using includes all the paragraph styles that you need, you should only need to apply character formatting.
- If you find that you need to create some new styles for a document, you may want to add them to the template that you're using.

Figure 8-5 When and how to use direct formatting

Other uses of the Styles pane

You've already learned how to open the Styles pane and how to use that pane to apply styles. Now, figure 8-6 shows some other uses of this pane.

First, you can use the drop-down list for a style to modify the style, as shown by the first procedure in this figure. This is an easy way to modify a style after you have applied direct formatting to a paragraph with that style.

If, for example, you update the Indented style in the last figure to the formatting for the selected paragraph, all of the paragraphs that have the Indented style are immediately updated. Also, the Indented + Right 0.5" style that was created by the direct formatting is removed from the Styles pane.

Second, if you want to create a new style, you can click the New Style button. Or, if you want to modify a style, you can click the Modify option in the drop-down list for the style. You'll learn more about creating and modifying styles in figures 8-8 and 8-9.

If you're like most Word users, those are the only other features of the Styles pane that you'll need. But if you're a power user, you can experiment with the other buttons and options in this pane. For instance, the Options button lets you determine which styles are shown in the Styles pane and what sequence they're sorted into. So, for example, you can show all of the styles in the document or just the ones in use.

A Style pane that shows the drop-down list for a style

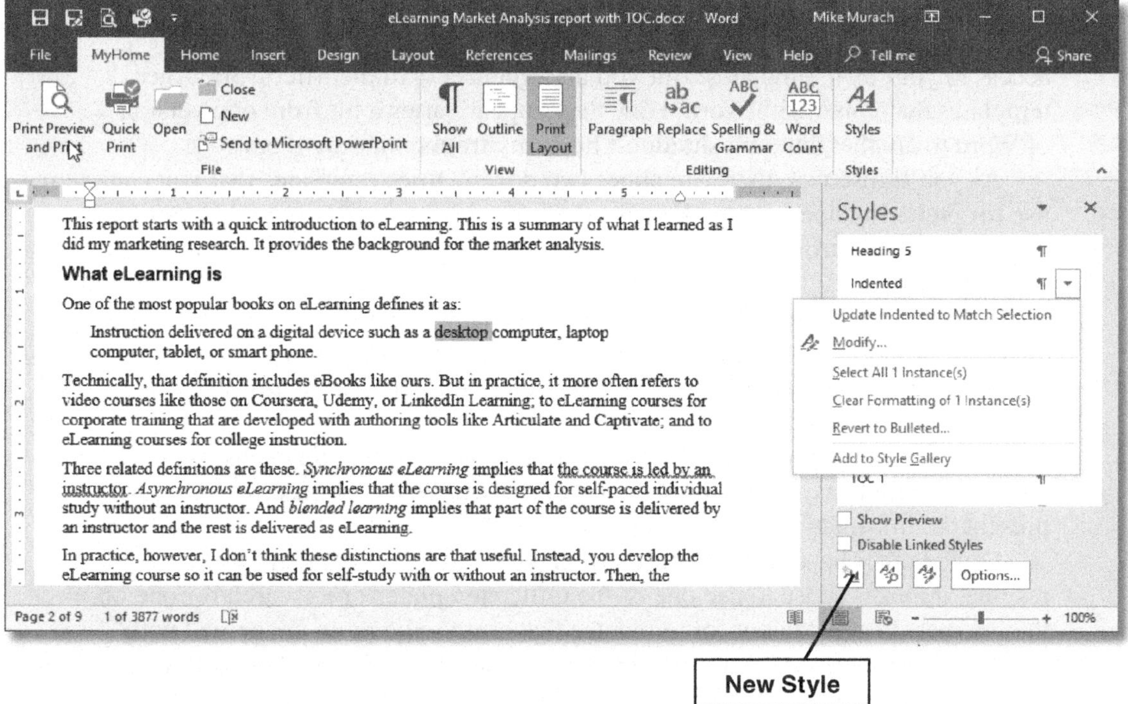

How to use the drop-down list to modify a style

- Move the insertion point into the paragraph that has the direct formatting that you want for the style.
- In the Styles pane, drop down the list for the style that you want to update. Then, select Update *style-name* to Match Selection where style-name is the name of the style. That will update all of the paragraphs that have that style applied to it.

How to open the Modify Style and Create New Style dialog boxes

- Click Modify in the drop-down list for a style to open the Modify Style dialog box (see figure 8-9).
- Click the New Style button to open the Create New Style dialog box (see figure 8-8).

Description

- If you're like most Word users, these are the other uses of the Styles pane that you're most likely to need. But if you're interested, you can experiment with all of the other buttons and options.

Figure 8-6 Other uses of the Styles pane

When and how to use the Office templates

When you start a new document with the File→New command, you get access to your own templates. But you also get access to the Microsoft Office templates that come with Word. How they appear varies a bit from one version of Word to another, but you shouldn't have any trouble finding them.

As you'll discover, there are about two dozen Office templates that you can use for preparing documents like letters, faxes, blog posts, flyers, and resumes. There's also a control in the New window that lets you search for other templates on the internet.

To step through the Office templates in the New window, you can click the Take a Tour icon. To illustrate, two of the templates on the tour are shown in figure 8-7. Then, to start a new document from one of the templates, you can click on its Create button.

In general, though, you're better off using your own templates instead of the Office templates. That way, your templates will be easy to use, and they will present the information just the way you want it.

On the other hand, if you don't have a template for a document like a resume, it makes sense to use one of the Office templates or to search for one on the internet. Just watch out for templates that seem to be more concerned with graphics than content because it's the content, not the graphics, that will make your documents successful. What you want is a template that helps you deliver your content as efficiently and as professionally as possible.

A report template provided by Word

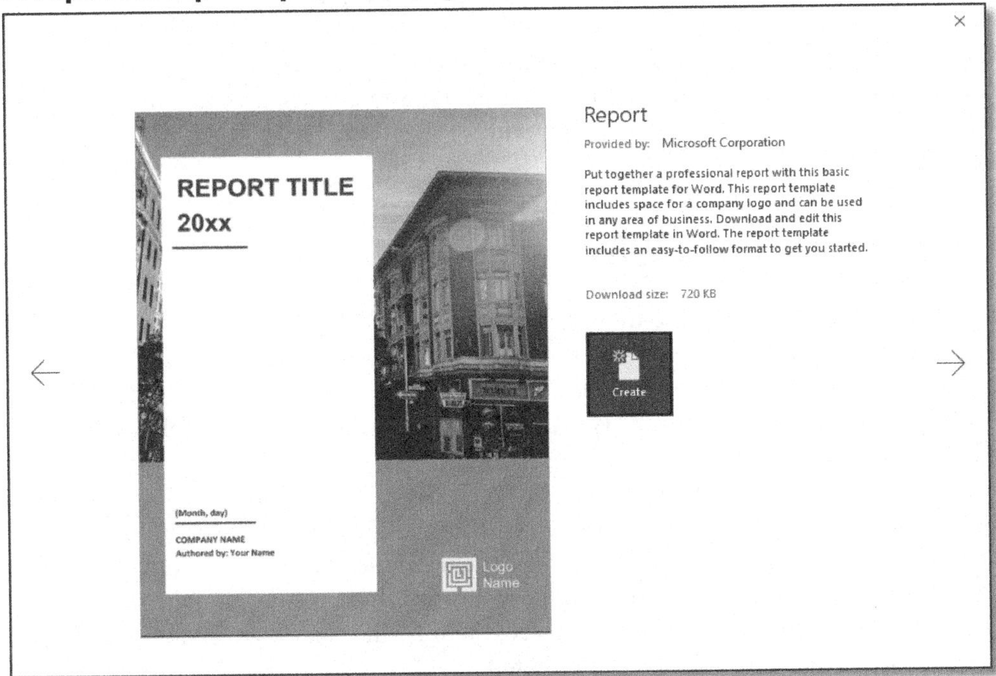

A resume template provided by Word

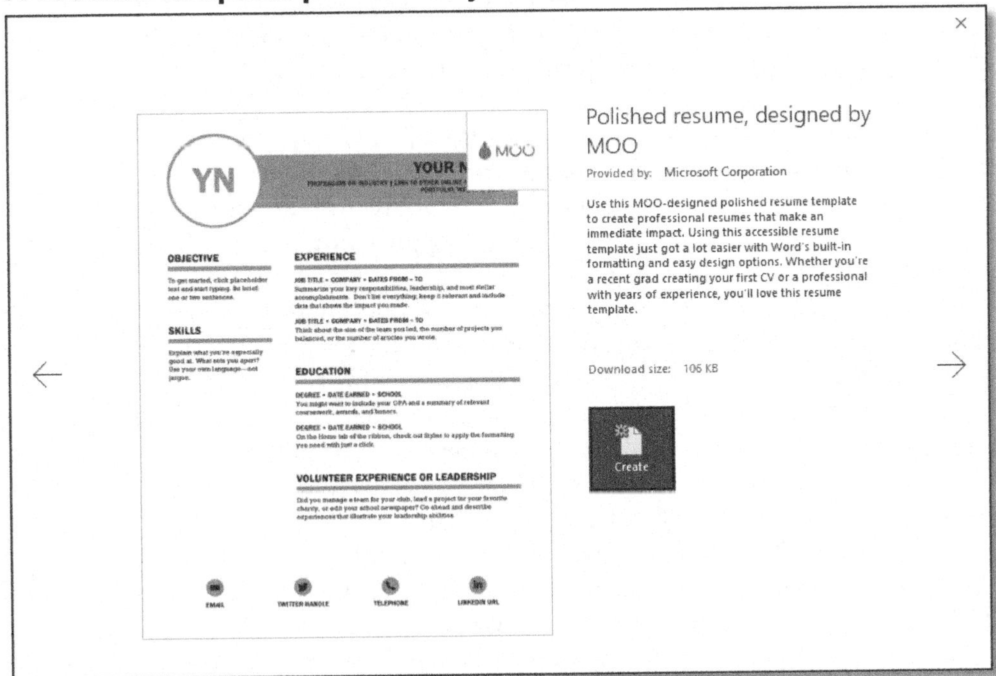

Description

- Word's Office templates are available when you select the File→New command. That includes templates for calendars, resumes, cover letters, brochures, and more.

Figure 8-7 When and how to use the Office templates

How to create and modify styles

With the skills you've just learned, you should be able to use existing templates and styles as you develop your documents. Sometimes, though, you will want to create new styles or modify existing styles. To do that, you can use the procedures that follow.

How to create a new style

To create a new style, you use the Create New Style dialog box shown in figure 8-8. To open this dialog box, you can click the New Style button in the Styles pane that's shown in figure 8-6. Then, you enter a name for the style and select Paragraph or Character as the style type. In this example, a paragraph style named Numbered will be created.

Next, you select the style that you want to base the new style on. If, for example, you want to create a new style that is like the Normal style except that it's indented, you can choose Normal as the base style. Then, you just need to modify the new style so it has the indentation you want. In this example, the base style is the Bulleted style so the Numbered style will be like that style except the paragraphs will be numbered instead of bulleted.

By default, the base style that's shown when this dialog box is opened will be the style for the paragraph that the insertion point is in. As a result, it makes sense to put the insertion point in the right paragraph before you open this dialog box. That way, you can also check the formatting of that style to make sure it's what you want.

The fourth entry for a new style is the style that should be applied to the paragraph that follows it when the user presses the Enter key to start a new paragraph. In this example, a Numbered paragraph will be followed by another Numbered paragraph. As a result, you will need to apply a different style to the first paragraph after the numbered list.

After you make those first four entries, you can use the controls in the Formatting group to modify the style. You can also use the drop-down list for the Format button to provide any of the other types of formatting.

When you're through with the formatting, you can uncheck the Add to the Styles Gallery box. If that box remains checked, the new style will be added to the Styles gallery in the Home tab. But since you'll be using the Styles pane to work with styles, you won't be using the Styles gallery.

Next, if you want the new style to be updated whenever the base style is changed, you can check the Automatically Update box. That can be useful, for example, if you want any changes to the base style to be applied to the new style.

Last, if you want the style added to the template that you're using as well as the current document, you should click the option button for New Documents Based on This Template. Otherwise, the new style will only be added to the document, not the template.

The Create New Style dialog box

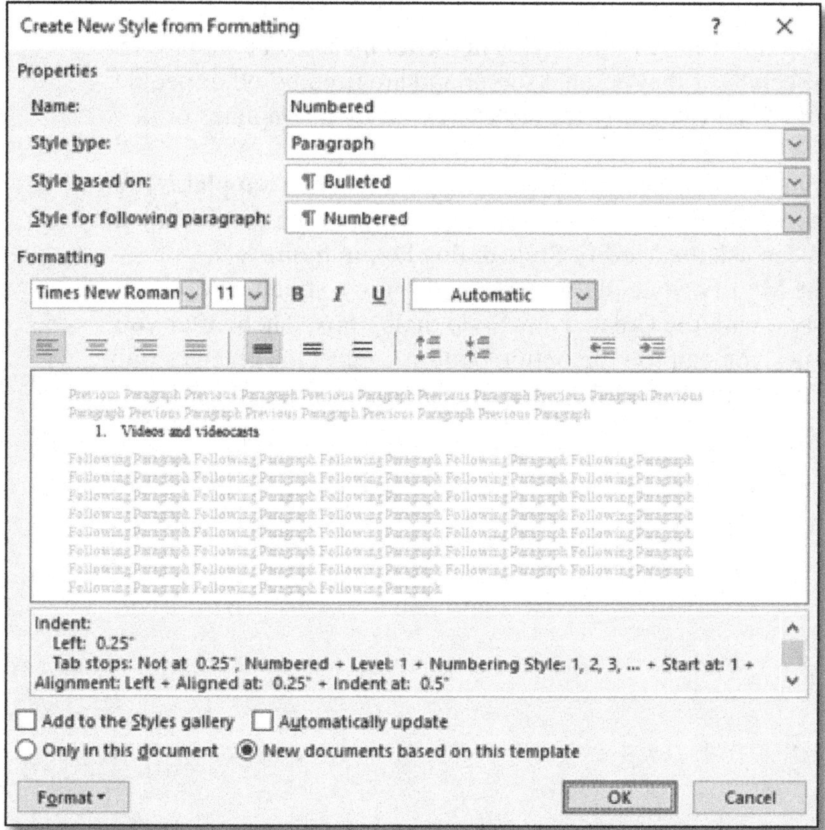

How to add a new paragraph style to a document

- Click on the New Style button in the Styles pane to open the Create New Style dialog box. Then, enter a name for the new style, and select Paragraph from the Style type list.
- By default, the style will be based upon the paragraph that the insertion point is in. If necessary, though, you can select the style that the new style should be based on. Then, select the style for the following paragraph.
- Use any of the controls in the Formatting group to change the font or paragraph formatting. To apply other formatting changes, drop down the Format list.
- If you want to stop the new style from being added to the Styles gallery, uncheck the Add to the Styles gallery box. And if you want the style to be automatically updated when the base style is changed, check the Automatically Update box.
- If you want to add the new style to the template that the document is based on, click the New Documents Based on This Template button. Otherwise, it will only be added to the document.

Figure 8-8 How to create a new style

How to modify a style

In general, you shouldn't modify the styles in the templates that are provided by your company, especially if they're on a server and are used by other people too. However, you may want to modify the styles in your own templates or in your documents.

To modify the style in a document, but not in the document's template, you can use the first procedure in figure 8-6. To modify a style in both the document and its template, you can use the Modify Style dialog box in figure 8-9.

When you use the Modify Style dialog box, modifying a style works much like creating a new style with the Create New Style dialog box. Then, after you make the modifications, you can use the option buttons to save the changes only in the current document or in the template that the document is based on.

The Modify Style dialog box

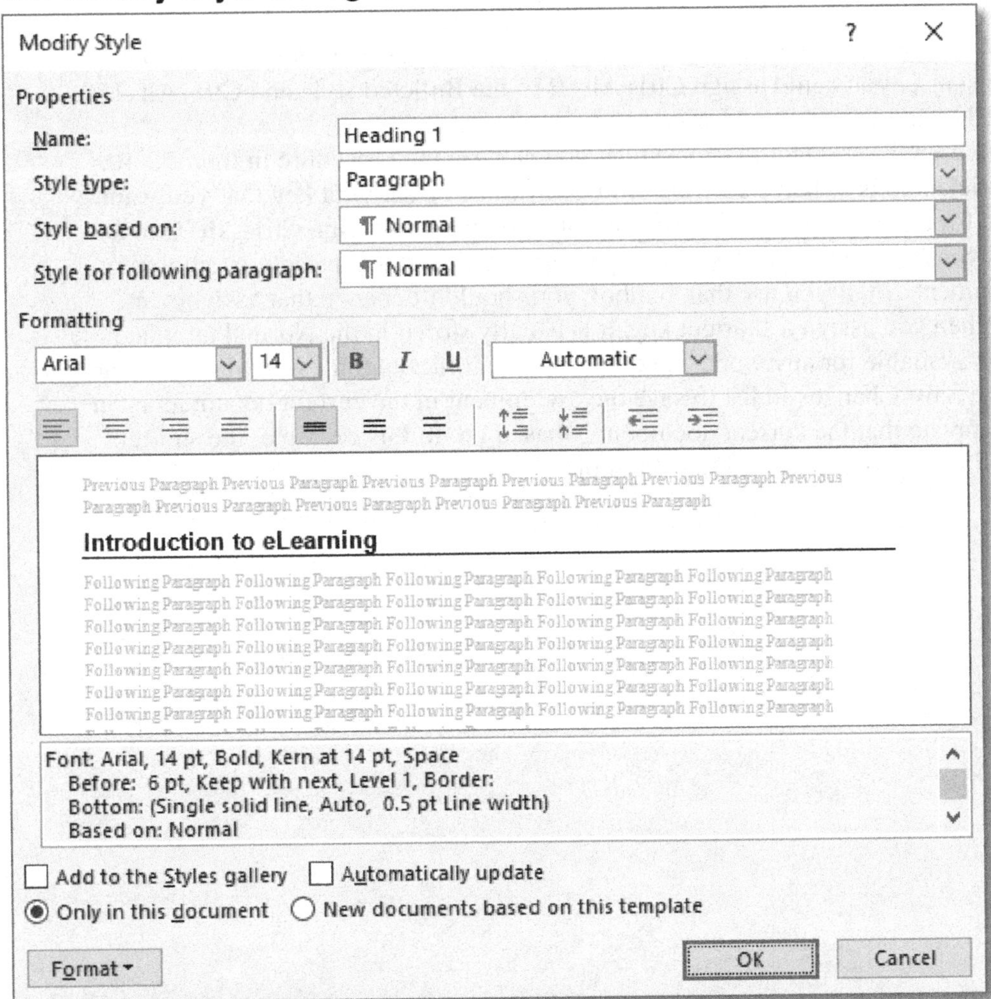

How to modify a style by using the Modify Style dialog box

- In the Styles pane, drop down the list for the style you want to modify. Then, select the Modify command.
- Use the Modify Style dialog box to modify the style. This works like the Create New Style dialog box in the previous figure.

Description

- Figure 8-6 shows a quick way to modify a style in a document, but that doesn't change the style in the template.
- If you want to modify a style in the related template, you need to use the Modify Style dialog box.

Figure 8-9 How to modify a style

How to assign a shortcut key to a style

When you create or modify a style, you can also assign a *shortcut key* to it. For instance, you could assign Ctrl+Alt+B to the Bulleted style and Ctrl+Alt+I to the Indented style.

To assign a shortcut key to a style, you can use the procedure in figure 8-10. The trick here is to make sure that you don't assign a shortcut key that you want to use for another function. If, for example, you want to assign Ctrl+Alt+C to a style, you'll find that the current assignment is to insert a copyright symbol into a document. So, if you use that symbol, you shouldn't change that assignment.

When you assign a shortcut key, it is usually stored in the Normal template so it is available for any work you do on your PC. If you don't want that, you can use the Save Changes In list to save the assignment in the current document or in the template that the current document is based on. In this example, the shortcut key will be assigned to the Report template.

The Customize Keyboard dialog box

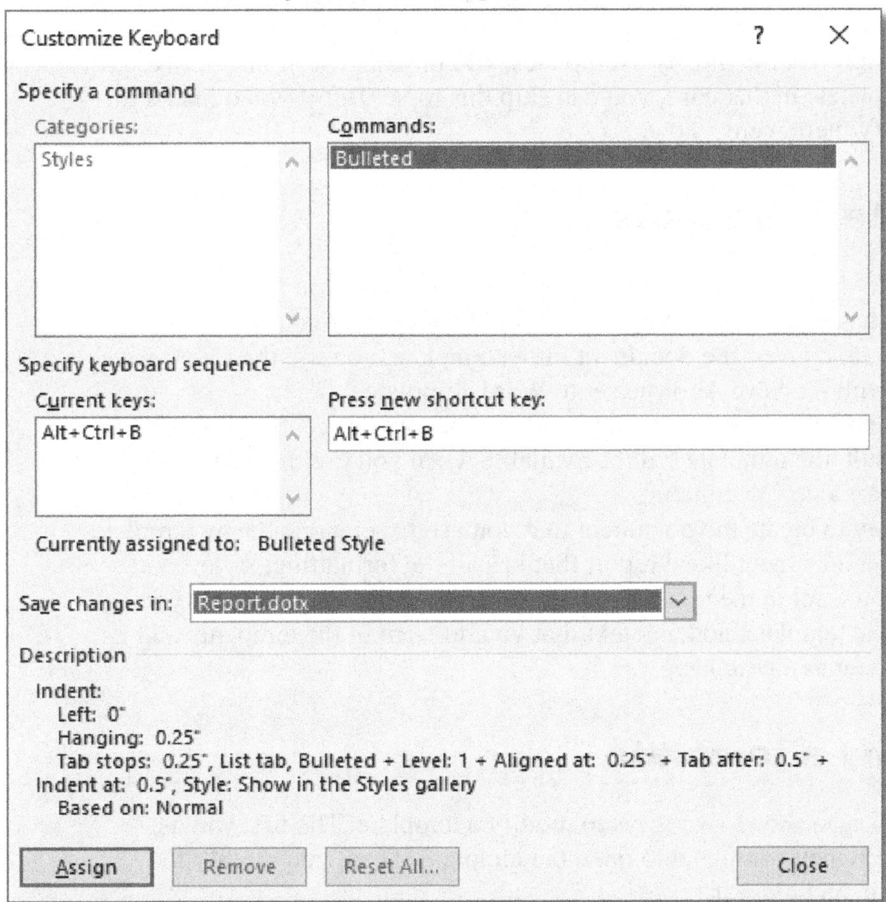

How to assign a shortcut key to a style

- In the Create New Style or Modify Style dialog box, click on the Format button and then on the Shortcut Key command. That will display the Customize Keyboard dialog box.
- Move the insertion point into the Press New Shortcut Key text box, and press the shortcut key you would like to use for the style. If that shortcut key is in use, its current assignment will be displayed in the Currently Assigned To textbox. Then, you can decide whether you want to choose a different shortcut key or change that assignment.
- Drop down the Save Changes In list, and select from the Normal template, the current template, or the current document.
- To complete the assignment, click the Assign button.

Description

- When you create or modify a style, you can also assign a *shortcut key* to it.

Figure 8-10 How to assign a shortcut key to a style

How to create or modify a template

If you're using your company's templates, you probably won't need to create or modify templates. In that case, you can skip this topic. But if you do need to create or modify them, here's how.

How to create a template

Figure 8-11 shows how to create a template. To do that, you first create a document that has all the formatting, styles, and features that you want in the template. Then, to convert the document into a template, you use the File→Save As command with the Save As Type set to Word Template.

That automatically saves the template in the Custom Office Templates folder. As a result, the template will be available when you use the File→New command to start a new document.

An easy way to create the document that you use for creating a new template is to start from a document like a report that has all the formatting, styles, and features that you want in the template. Then, you just delete any text that you don't want in the template, add any text that you do want in the template, and save the document as a template.

How to modify a template

Figure 8-11 also shows two ways to modify a template. The first way is to use the File→Open command to open the template. Then, you modify the template, save it, and close it.

The second way is to use the Modify Style dialog box to change the styles in a document, and to save those changes in the template that the document is based upon. This works well when you're just making minor changes to the styles.

The Save As dialog box for creating a new template

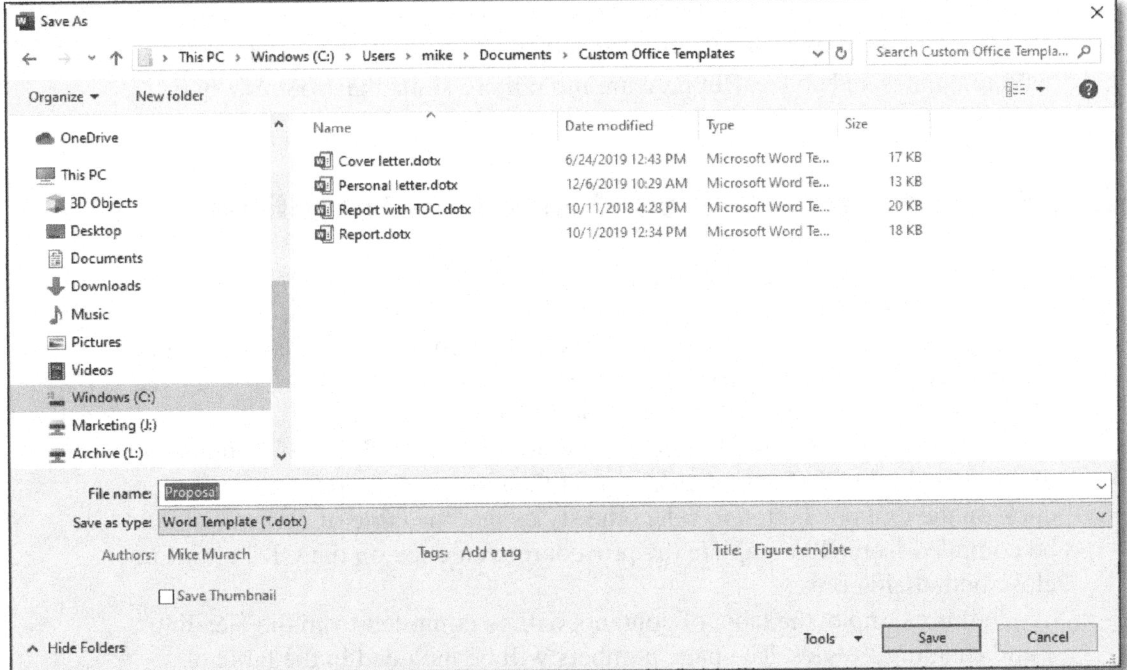

How to create a new template

- Create a document that has all the styles and features that you want in the new template.
- Use the File→Save As command to save the document with Save as Type set to Word Template.

How to modify a template directly

- Use the File→Open command to find and open the template.
- Modify the template.
- Save and close the template.

How to modify a template indirectly

- Use the Modify Style dialog box to modify a style, and turn on the option for New Documents Based on this Template. Then, click OK to close the dialog box.

Description

- By default, when you create a new template, it will be saved in the Custom Office Templates folder. Then, when you issue the File→New command to start a new document, you'll be able to find the template in the New window.

Figure 8-11 How to create or modify a template

Related skills

At this point, you may already have all the skills that you need for working with templates and styles. But here are three more skills that you may find useful.

How to add a table of contents to a template

For a longer document like a report or proposal, it often helps to include a *table of contents* on the first page of the document. For that reason, it also makes sense to include a table of contents in the templates for creating longer documents. That's why figure 8-12 shows how to do that.

To start, you move the insertion point to where you want the table of contents. Then, you use the References tab to display the Table of Contents dialog box. In that dialog box, you can set the options that you want, and then click on the Options button to select the styles that the table of contents should be compiled from. To complete the procedure, you click on the OK buttons to close both dialog boxes.

In this example, the table of contents will be compiled from the Heading 1 and Heading 2 styles. The page numbers will be included in the table of contents. They will be right aligned. And they will have a dot leader. You can see how that looks in the table of contents in the document.

When the table of contents is inserted into the template, the two levels in the table will be formatted by the TOC 1 and TOC 2 styles. These are built-in Word styles, but you can modify them if necessary. In this example, these styles have been modified so they're indented from the left and so the TOC 1 style is bold but the TOC 2 style isn't.

A table of contents on the first page of a document

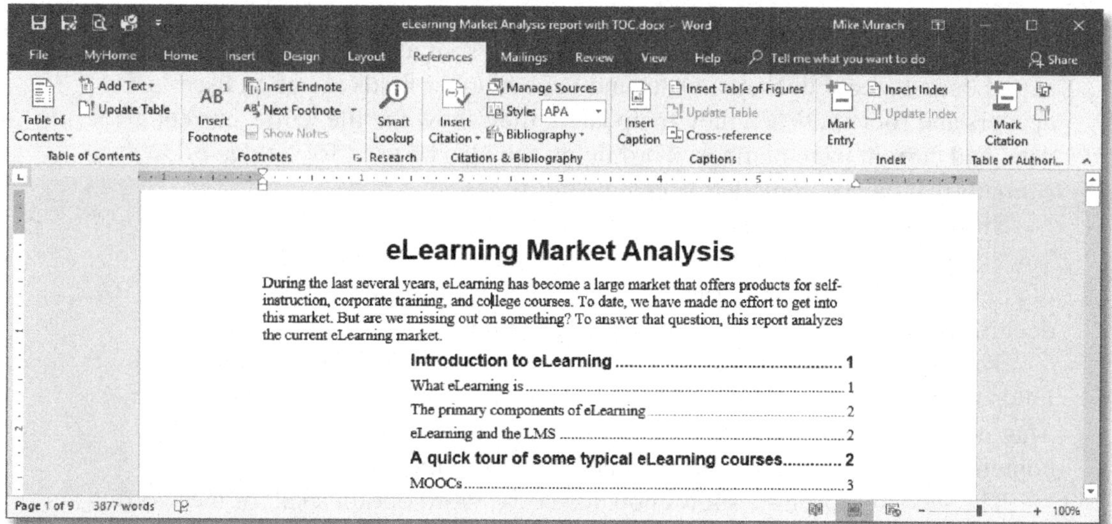

The dialog boxes for adding a table of contents

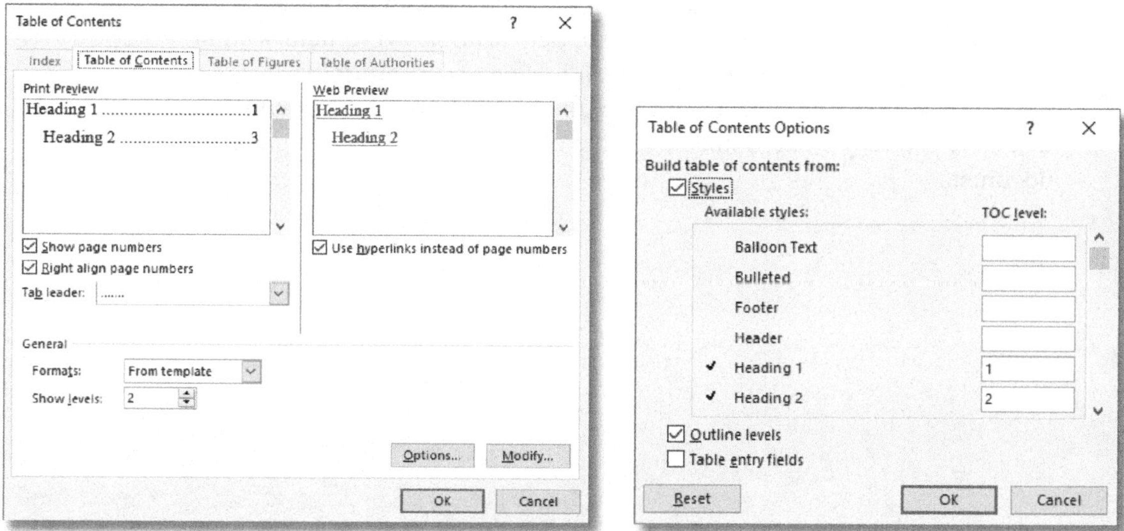

Procedure

- Move the insertion point to where you want the table of contents (TOC) inserted into the template. Then, use the References tab to drop down the Table of Contents list, and select Custom Table of Contents to display the Table of Contents dialog box.
- In the Table of Contents dialog box, set the options, click the Options button, set the options in that dialog box, and click OK to close each of the open dialog boxes. That will insert the TOC into the template.
- If necessary, you can then modify the TOC 1 and TOC 2 styles in the template to get the TOC to look the way you want it.

Figure 8-12 How to add a table of contents to a template

How to update Word fields

A *Word field* is a feature that can be updated so it always has the current data. For instance, FileName, Date, and Page number fields are often used in headers and footers. When they're updated, they show the file name, current date, and current page number. Word fields can also be used for a *table of contents* that can be compiled from heading styles.

Since Word fields are commonly used in templates, figure 8-13 shows how to update them. Here, the first example contains a Word field for a table of contents. The second example contains three Word fields in the footer of a document: file name, current date, and page number.

The first procedure in this figure shows how to set the Word options so the fields in your documents are shaded whenever they're selected or all of the time. That makes it easier to identify the Word fields. In the examples, the table of contents and the file name field are shaded because they're selected.

The second procedure shows how to set the Word option so all of the Word fields are updated whenever the document is printed. That way you can be sure the data is accurate in the printed document. If the fields aren't updated when the document is printed, the printed document could contain inaccurate data.

The third procedure shows how to update a single Word field without printing the document. That can be useful when you're editing and you want to see what the current data is, but you don't want to print the document. For instance, you may want to update a table of contents before you print the document.

The table of contents field in a report

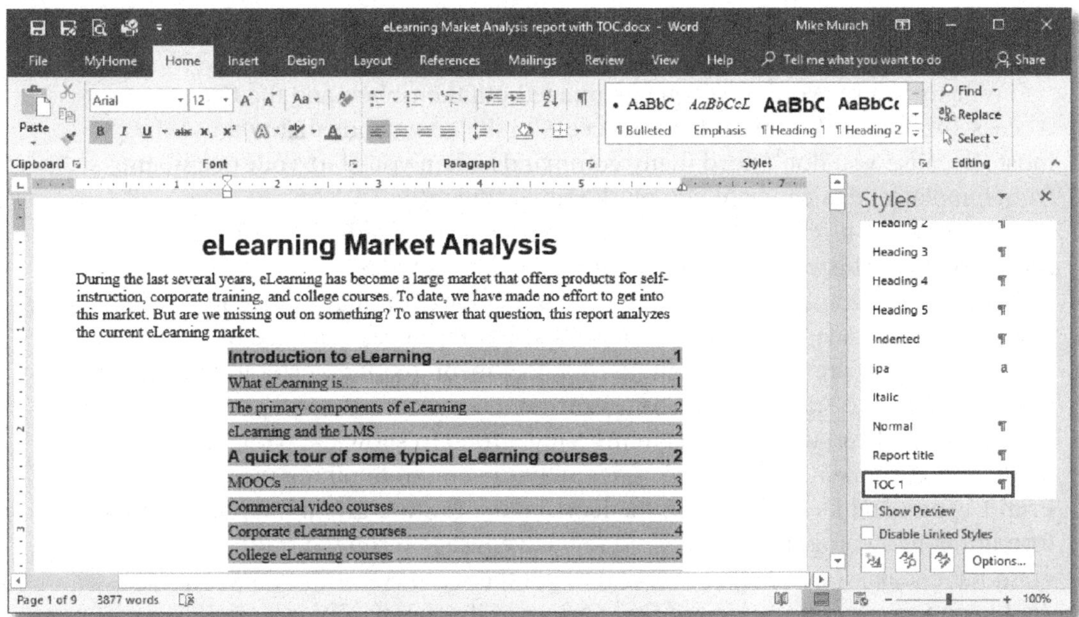

The FileName field in the footer of a report

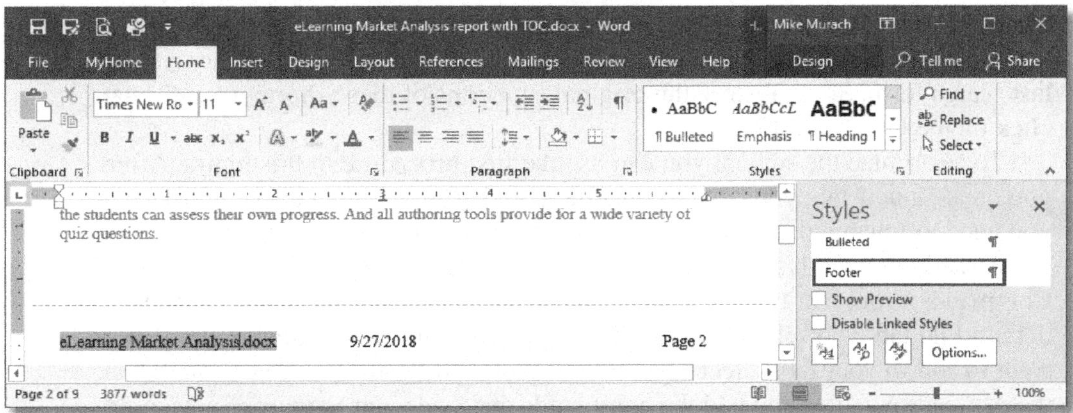

How to shade the Word fields

- Select File→Options, click on Advanced, scroll to Show Document Content, and set Field Shading to either When Selected or Always.

How to update the Word fields whenever a document is printed

- Select File→Options, click on Display, scroll to Print group, and check Update Fields Before Printing. Then, the fields will be updated every time you print the document.

How to update a Word field without printing the document

- Right-click in the Word field, and select Update Field from the shortcut menu.

Figure 8-13 How to update Word fields

How to create a custom ribbon tab or Quick Access Toolbar

When you use Word, you can customize both the ribbon and the Quick Access Toolbar so they provide easy access to the commands that you use the most. Because you don't need many commands when you start your documents from templates, this can help you work more efficiently.

In figure 8-14, for example, you can see the customized tab that I created for the ribbon. This tab is named MyHome, and it provides the commands that I use more than 90 percent of the time. That includes commands that are copied from the File, Home, View, and Review tabs. For instance, I can use the icons in my File group to preview the printing of a document, to quick print it, to open or close a file, and to start a new document from a template.

Similarly, I can use the icons in my View group to show the paragraph marks in a document or to display a document in any of the three views: Outline, Print Layout, and Draft. I can use the icons in the Editing group to apply direct formatting to a paragraph, to find and replace entries, to run the spelling and grammar checker (see chapter 10), and to get the word count for a document. And I can use the one icon in the Style group to display the Styles window.

But note that the Editing group also includes a Go To Next Page icon. I added that icon to the group because the normal functions of the Ctrl+PgUp and Ctrl+PgDn shortcut keys aren't restored when the Find and Replace is finished. Instead, these keys continue to search for the next or previous occurrence of the last search term. So, to restore the original functions of these shortcut keys, I can click on the Go To Next Page icon.

To customize the ribbon, you can use the first procedure in this figure. After you create a new tab, you need to rename it, and after you create a new group, you need to rename it. Then, you can add commands to the new groups.

To add commands to a group, you first select an item from the Choose Commands From list. If, for example, you want to choose commands from the default File tab, you select File Tab. Then, you can find the command that you want to add to your tab, select it, select the group you want to add it to, and click the Add button. After you add the commands that you want to your groups, you can re-sequence the tabs, groups, or commands by selecting one and clicking the up or down arrow.

If you try this, you'll see that it takes some trial and error to get a custom tab the way you want it. But once you get it right, it becomes a lot easier to start the commands that you use the most.

Then, to take this a step further, you can customize the Quick Access Toolbar. This is the toolbar that's right above the Word ribbon with the small icons on it. In this figure, this toolbar has icons for the Save and Save As commands plus three others.

To customize this toolbar, you can use the second procedure in this figure. To start, you can drop down the list from the right side of the toolbar. Then, you can check any of the common commands in this list to add them to this toolbar. And if you want to add other commands, you can click More Commands to display the Word Options dialog box again, but in a simpler form.

A custom MyHome tab and Quick Access Toolbar

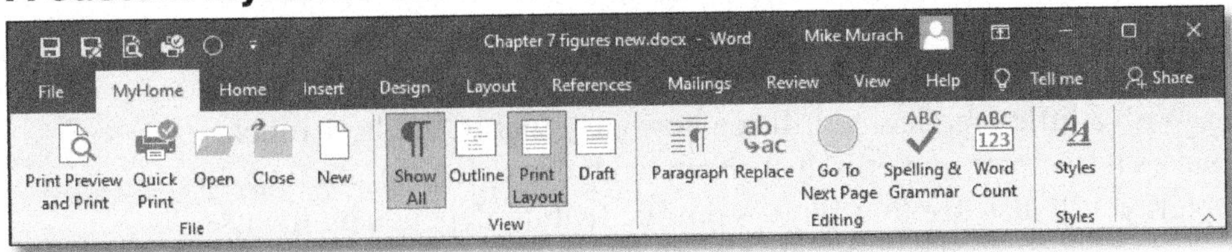

The Word Options dialog box for creating the MyHome tab

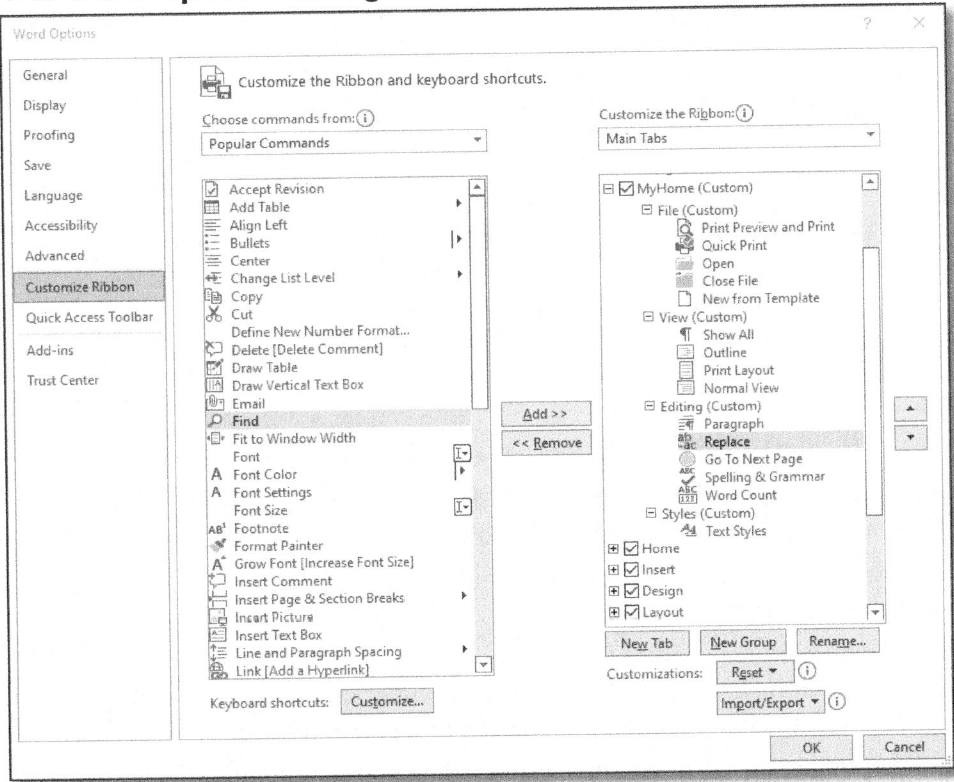

How to create a custom tab in the Word ribbon

- Right-click in the ribbon, and select Customize Ribbon. Then, in the Word Options dialog box, use the buttons to start your own tab or group within that tab.
- To add a command, drop down the Choose Commands From list, and choose a source. Then, select the command, select the group in the right panel, and click the Add button.
- To re-sequence a group or a command, select it and click the up or down button.

How to customize the Quick Access Toolbar

- Click the drop-down arrow on the right side of the Toolbar. Then, check the commands you want in it. Or click More Commands to open the Word Options dialog box.

Figure 8-14 How to create a custom ribbon tab or Quick Access Toolbar

Perspective

I hope by now you're sold on the idea of using templates and styles for your writing. With a little practice, you'll see how easy that is and how much easier that makes it to create and edit documents with consistent formatting. In the next chapter, you'll also see that Word's outline feature depends on the use of styles.

If you haven't already downloaded the files for this book, appendix A shows how. That download includes the two report templates that are illustrated in this chapter so you can start new documents from them and use their styles.

Terms

template
style
built-in style
paragraph style
character style
shortcut key
direct formatting
indirect formatting
Word field
table of contents

Summary

- A *template* is a document that's used to start other documents. A template includes *styles* that can be used to format the paragraphs in a document.
- When a *paragraph style* is applied to a paragraph, it formats both the paragraph and the characters within the paragraph.
- To apply a paragraph style, you can use the Styles pane or a *shortcut key*. In addition, a style is automatically applied to the next paragraph when you press the Enter key with the insertion point at the end of a paragraph.
- You can use *direct formatting* when you just want to change the formatting for a paragraph or two without creating a new style for it.
- Besides applying styles, you can use the Styles pane to open the Modify Style or Create New Style dialog boxes. You can also use the drop-down list for a style to update the style so it has the formatting of the selected paragraph.
- When you use the File→New command to start a new document, you get access to your own templates as well as the Microsoft Office templates.
- To create a template, you first create a document that has all the styles and features that you want in the template. Then, you use the File→Save As command to save the document as a template.
- A *Word field* can be used for data like a date or a *table of contents*. To update a Word field, you can use the shortcut menu of the field, or you can set the options so all Word fields are updated when a document is printed.
- To make it easier to access the Word commands that you use the most, you can create a custom ribbon tab and add commands to the Quick Access Toolbar.

9

How to use Word's outline feature

The outline feature is another feature that every writer should use. It is especially useful for planning the headings and subheadings that you're going to use in a document. But you can also use it to plan your activities, organize your research, and more. Besides that, this feature is remarkably easy to use.

How to create and use an outline

An *outline* is a series of paragraphs that have heading styles applied to them. The structure of the outline is shown by the indentation of these paragraphs. For instance, a Heading 1 paragraph isn't indented, a Heading 2 paragraph is indented one tab, and so on.

In figure 9-1, for example, you can see an outline that has two heading levels. This is the start of a heading plan for a report on eLearning.

How to start an outline

To start a Word outline, you start a new document from any template and switch to Outline view using the technique in figure 9-1. We recommend, however, that you start your outlines from a template that has a header or footer that shows the file name and date as well as the formatting that you want for the Heading 1, Heading 2, and Heading 3 styles.

When you switch to Outline view, you will automatically be entering paragraphs that have the heading styles applied to them, starting with a Heading 1 style. Then, when you press the Enter key at the end of a paragraph, another paragraph at the same level is started.

To lower the level of the style that's applied to a paragraph, you can press the Tab key. This is known as *demoting* a heading. Similarly, you can press the Shift+Tab shortcut key to *promote* the level of a heading. By using these keys, you can enter an outline into Outline view using the keyboard alone.

The plus and minus signs before the headings in an outline indicate whether or not a heading has subheadings. Although you can have up to nine heading levels in an outline, you should rarely need more than three.

As you can see in this example, the styles area pane on the left side of the screen shows which style has been applied to each paragraph. That's useful whenever you're having trouble with your formatting and want to see which style has been applied to each paragraph. That's why this figure also shows how to display this pane.

The good news is that this pane isn't displayed in Print Layout view, but it is displayed in Outline and Draft view. That means that if you need this pane, you can access it just by changing views. But when you don't need it, you can work in Print Layout view, which you'll do most of the time.

The start of an outline in Outline view

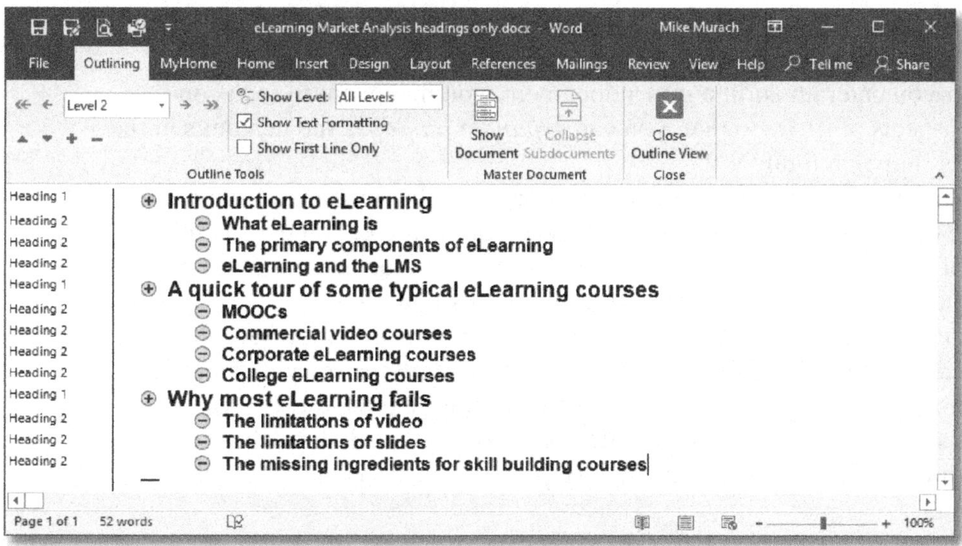

How to switch to and from Outline view

- To switch to Outline view, click on the Outline icon in the View tab. In Outline view, the Outlining tab is displayed.
- To switch back to the view you were in, click on Close Outline View in the Outlining tab.

How to enter an outline in Outline view

- When you switch to Outline view and start typing, the first heading that you type will be a level-1 heading that has the Heading 1 style applied to it.
- When you press the Enter key at the end of a line, another line is created at the same heading level.
- To *demote* (lower) or *promote* (raise) the level of a heading, you can press the Tab key or the Shift+Tab shortcut key.

How to display the style area pane to the left of the outline

- Use the File→Options command to open the Options dialog box. Next, click Advanced to go to the advanced options, and scroll down to the Display group. Then, set the Style Area Pane Width to 1 inch. This pane will now be displayed in Draft and Outline views, but not in Print Layout View.

Description

- An outline can have nine levels of headings that have the Heading 1 through Heading 9 styles applied to them. The plus and minus signs before the headings show whether the headings have subordinate levels.
- We recommend that you always start an outline from a template that has the styles and features that you need, including a footer that identifies the document.

Figure 9-1 How to start an outline

How to expand and collapse the headings in an outline

After you enter an outline into a document, you may want to focus on specific aspects of it. To do that, you can *expand* or *collapse* the headings in the outline as shown in figure 9-2.

To start, you can use the Show Level drop-down list in the Outlining tab to show the headings at a specific heading level. For example, you can set this to Level 1 if you want to review just the level-1 headings for sequence and consistency. This is illustrated by the first example in this figure.

Then, you can expand one level-1 heading at a time to review its subheadings for sequence and consistency. This is illustrated by the second example in this figure, which has just one of the level-1 headings expanded. To expand or collapse a heading, you can double-click on its plus sign, click on the Expand or Collapse button in the Outlining tab, or press the plus or minus key on the 10-key pad.

An outline that shows just the level-1 headings

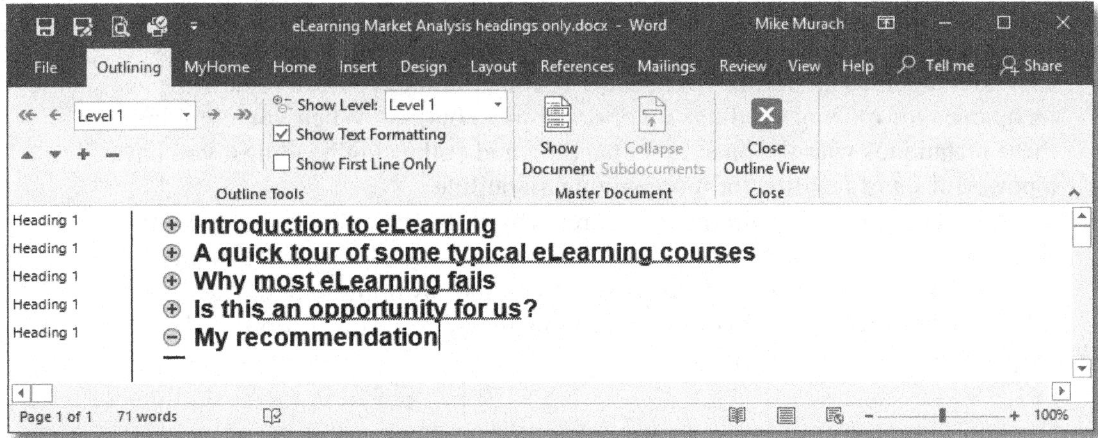

The outline after one of the level-1 headings has been expanded

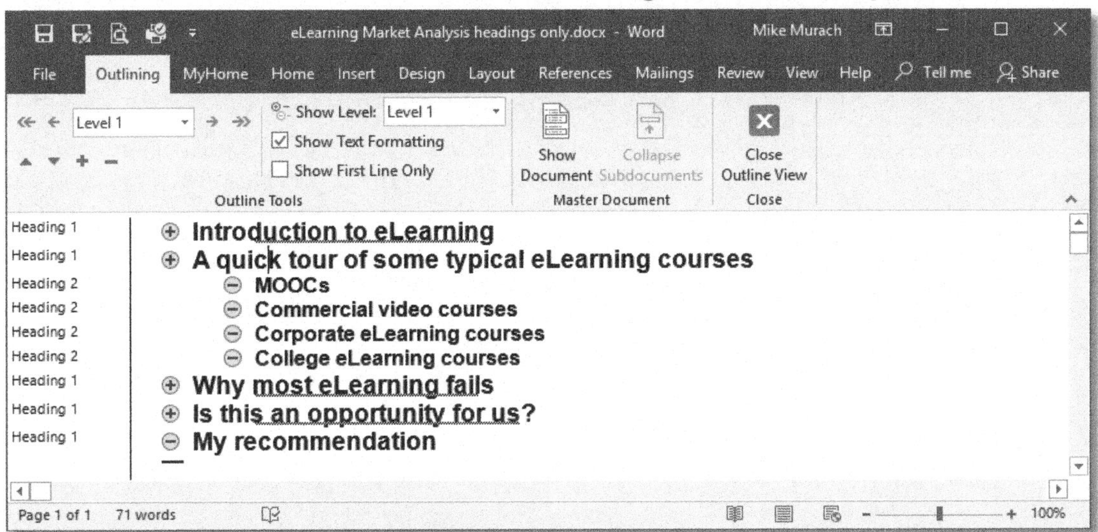

How to use the mouse to expand or collapse headings

- To display just the headings at and above a specific heading level, select a level from the Show Level list in the Outlining tab.
- To *expand* (display) or *collapse* (hide) the subheadings or text for a heading, click in the heading and click the Expand (plus) or Collapse (minus) button in the Outlining tab. Or, double-click on the plus or minus sign at the start of a heading.
- To display the headings with or without formatting, check the Show Text Formatting box in the Outlining tab.

How to use the keyboard to expand or collapse headings

- To expand or collapse the subheadings or text for a heading, press the + or – key on the 10-key pad.

Figure 9-2 How to expand and collapse the headings in an outline

How to reorganize an outline

One of the benefits that you get from using the outline feature is that it's easy to reorganize an outline. In figure 9-3, for example, you can review the techniques for moving headings up or down in an outline. When you combine these techniques with the ones for expanding and collapsing headings, you have a powerful set of features for reorganizing an outline.

To get comfortable with these features, you just need to experiment with them for a while. If you like working with a mouse, try all the controls in the Outlining tab to see what each one does. If you like working with the keyboard, try all the keystroke combinations to see how they work. You'll soon see how easy it is to work with an outline.

A document in Outline view with a heading and its subheadings selected

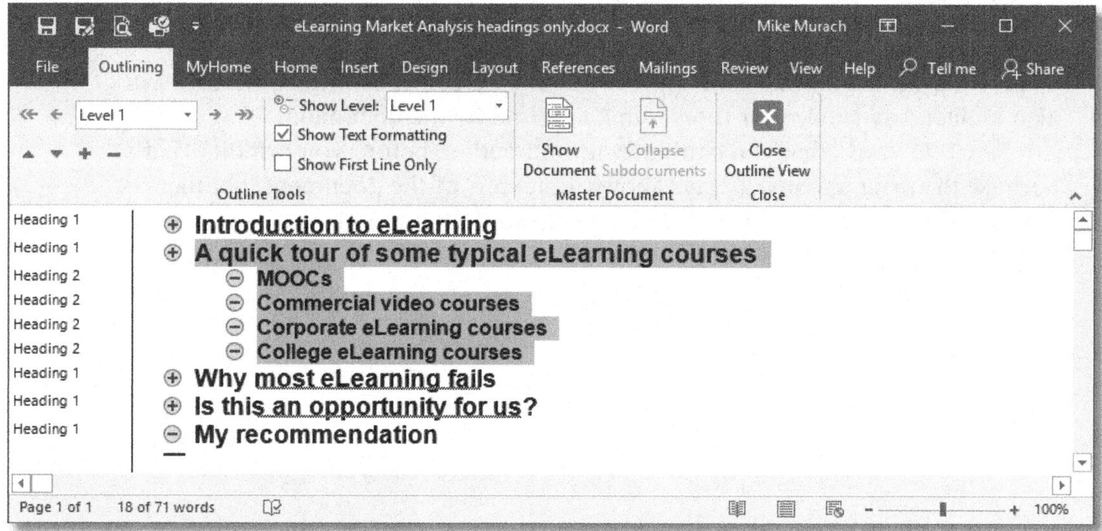

How to use the Outlining tab to work with headings

- To move a heading up or down in the outline, click on the Move Up (up arrow) or Move Down (down arrow) button. To promote or demote a heading, click on the Promote (left arrow) or Demote (right arrow) button.
- To raise a heading to level 1, click on the Promote to Heading 1 (double left arrow) button. To demote a heading to body text, click on the Demote to Body Text (double right arrow) button.
- To apply a specific heading level to a paragraph, use the drop-down list that's between the demote and promote buttons.

How to use the keyboard to work with headings

- To promote or demote a heading, press Shift+Tab or Tab. Or, press Alt+Shift+Left-arrow or Alt+Shift+Right-arrow.
- To move a heading up or down, press Alt+Shift+Up-arrow or Alt+Shift+Down-arrow.

How to use the mouse to work with headings

- To select a heading and all of its subheadings and text, click on its plus or minus sign. To select one or more headings, drag the mouse over them.
- To promote or demote a selection, drag it to the left or right. To move a selection, drag it to its new location or cut and paste it into its new location.

Figure 9-3 How to reorganize an outline

How to print an outline

To print an outline, just print the document while you're in Outline view. This will print the outline as it appears on your screen. The printed outline will also include any headers or footers that are used by the document.

Note, however, that you can't preview an outline before you print it. That's because the print preview always shows a preview of the document, not the outline. This is illustrated by the first screenshot in figure 9-4.

This is the way an outline is shown in print preview

Introduction to eLearning
What eLearning is
The primary components of eLearning
eLearning and the LMS
A quick tour of some typical eLearning courses
MOOCs
Commercial video courses
Corporate eLearning courses
College eLearning courses
Why most eLearning fails
The limitations of video
The limitations of slides
The missing ingredients for skill building courses
Is this an opportunity for us?
The competition
Our product possibilities
The sales potential
The development costs
My recommendation

But this is the way the outline is printed

Introduction to eLearning
    What eLearning is
    The primary components of eLearning
    eLearning and the LMS
A quick tour of some typical eLearning courses
    MOOCs
    Commercial video courses
    Corporate eLearning courses
    College eLearning courses
Why most eLearning fails
    The limitations of video
    The limitations of slides
    The missing ingredients
Is this an opportunity for us?
    The competition
    Our product possibilities
    The sales potential
    The development costs
My recommendation

How to print an outline that's shown in Outline view

- To print whatever is shown in Outline view, use your normal technique for printing.
- Just realize that the print preview will show the entire document including subheadings and text paragraphs that aren't shown in Outline view. But that isn't what gets printed.

Figure 9-4 How to print an outline

Other uses of the outline feature

If you take our recommendations, you will start your documents from the templates that are designed for them. Then, you will use the outline feature to develop the heading plans for your documents.

Please note, however, that the outline feature can also be useful while you're editing. And it works well for other types of planning and organization too.

How to use Outline view as you write and edit

As you write and edit a document, you may discover that you want to do some reorganization of the original heading plan. Although you could do that in Print Layout view by cutting and pasting, it's often more efficient to switch to Outline view and do the reorganization there.

For instance, figure 9-5 shows a document in Outline view. At this point, it's easy to reorganize the document by first collapsing any headings or subheadings that you want to move and then moving the headings or subheadings to new locations.

When you're done, you can return to Print Layout view. There, the cursor will be on the same heading or paragraph that it was in Outline view.

A document in Outline view

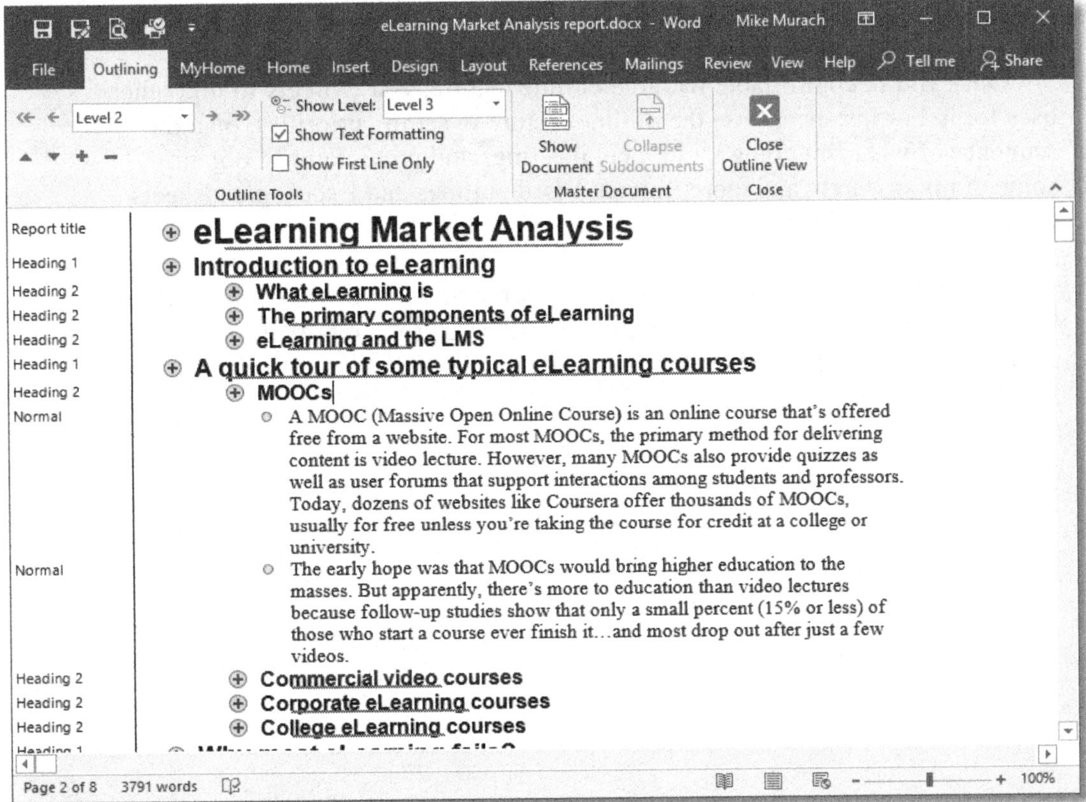

Description

- As you write and edit a document, you may decide to add, relocate, or delete topics and subtopics. Often, the easiest way to do that is to switch to Outline view and work with the headings and subheadings in that view.
- To show just the headings (not the text), use the Show Level drop-down list in the Outlining tab to set the number of heading levels that you want to show.
- When you delete or relocate a heading with collapsed text, the text is deleted or relocated along with the heading.
- When you're through with the changes, you can switch to Print Layout view. There, the cursor will be on the same heading or text paragraph that it was on in Outline view.

Figure 9-5 How to use Outline view as you write and edit

How to use Outline view to plan and organize your work

Once you're comfortable with the outline feature, you're likely to find other uses for it. For instance, I use the outline feature to set my priorities, plan the sequence of what I'm going to do, plan meetings and presentations, organize some of my research, and more. In fact, Word outlines and Excel spreadsheets are the primary tools that I use for business planning.

Figure 9-6 illustrates just two of these uses. Here, the first is the start of an outline that's used to organize and prioritize what needs to be done. The second is the start of an outline that summarizes the research for a report. That type of summary can often be used as the start of the heading plan for a document.

An outline for planning your priorities

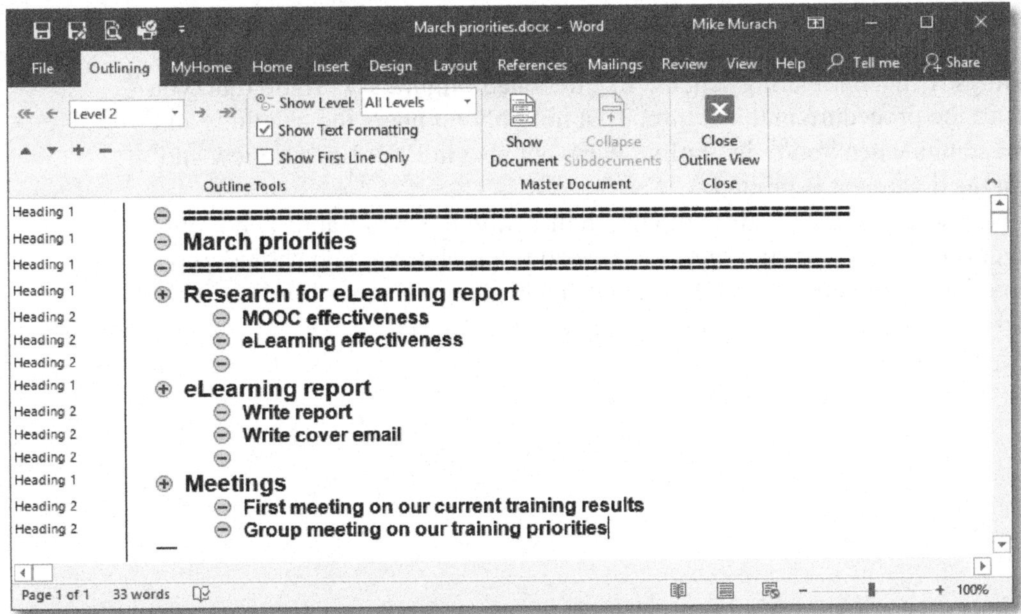

An outline for organizing your research

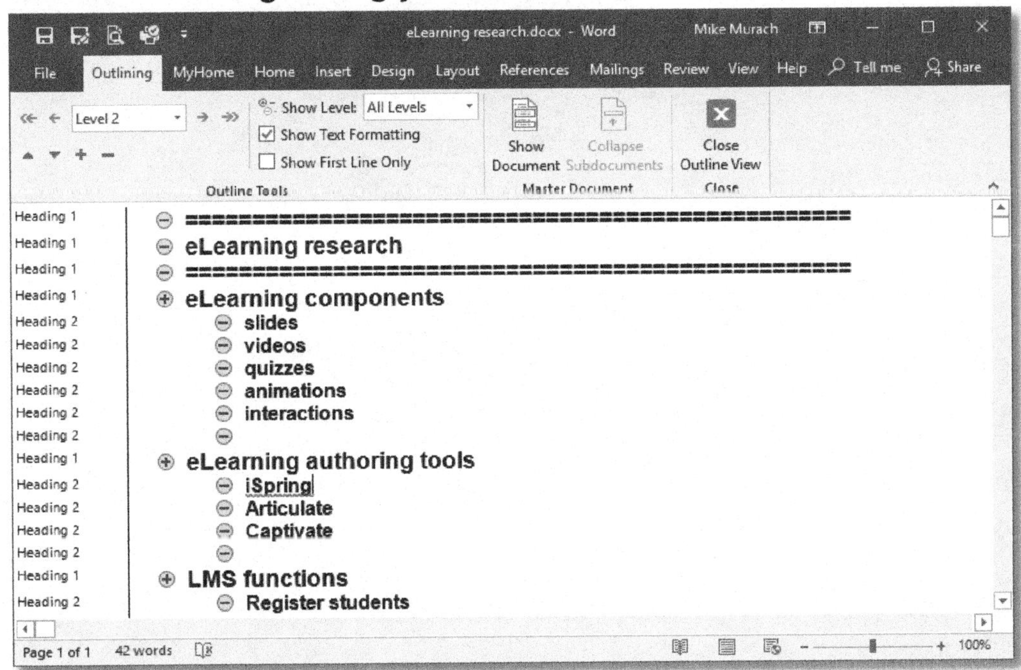

Description

- In addition to their use when you're writing and editing, Word outlines can be used to plan your priorities and organize your research.

Figure 9-6 How to use Outline view to plan and organize your work

When and how to number the headings

For some types of technical writing, you may be required to number the headings with a numbering scheme like the one in figure 9-7. To do that, you can use the procedure in this figure. That not only numbers the headings and subheadings when you're in Outline view, but also in Print Layout view and when the document is printed.

For most business writing, though, numbering the paragraphs is not only unnecessary but also distracting. So, instead of numbering, you should use formatting to show the difference between a heading and a subheading. That's a better way to show the structure of your topics and subtopics.

How numbered headings look in Outline view

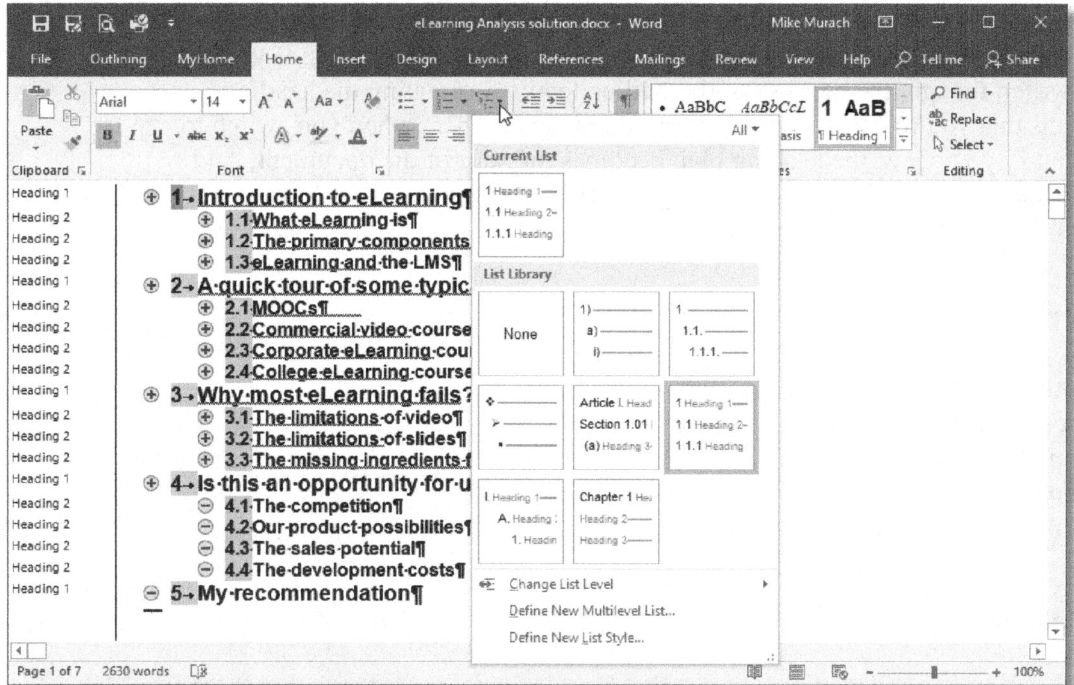

How the headings are formatted in Print Layout view

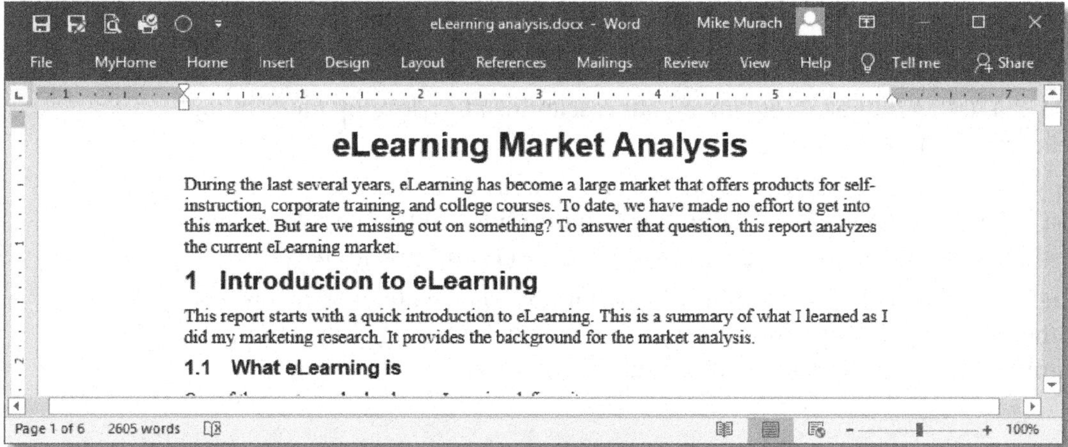

How to number the headings and subheadings

- While you're in Outline view, switch to the Home tab in the ribbon, and use the drop-down Multilevel List to select a numbering scheme.

Description

- For some types of technical writing, you may need to number the headings. But for most documents, you shouldn't number them.

Figure 9-7 When and how to number the headings

Perspective

As simple as they are, outlines can be used for many purposes. If you're a writer, though, the best use of the outline feature is creating the heading plans that you're going to use for your documents. Then, when you switch to Print Layout view, the heading plan becomes the start of the document. And whenever you need to do some reorganization, you can just switch back to Outline view.

Terms

outline
promote a heading
demote a heading
expand a heading
collapse a heading

Summary

- An *outline* is a series of paragraphs that have heading styles applied to them. In Outline view, the heading levels are identified by successive levels of indentation.
- Outline view makes it easy to *promote* or *demote* a heading. It also makes it easy to *expand* or *collapse* headings.
- You can use an outline to organize your research and to develop the heading plan for a document.
- As you write and edit a document, you can switch to Outline view whenever you want to reorganize the headings and subheadings. When you're through working with the outline, you can switch back to Print Layout view.
- If necessary, you can use Outline view to number the headings in an outline. Although that's required for some types of technical writing, you shouldn't number the headings unless that is required.

10

How to use Word's spelling and grammar checker

No matter how careful you are, the spelling and grammar checker will usually help you find an error or two in each document. That's why all writers should try to get the most from this feature. In this chapter, you'll learn how to do that.

You should know, however, that the spelling and grammar checker often reports errors that aren't really errors. As a result, you need to be able to tell when a suggested change is correct. So, in case you need it, this chapter ends with a quick reference for grammar. It will help you recognize the correct suggestions and skip the ones that aren't.

Two ways to use the spelling and grammar checker

The *spelling and grammar checker* can check for errors in two ways. First, it can check for errors as you type and highlight any that it finds. Second, it can check for errors when you run the spelling and grammar checker and present them to you one at a time.

How to fix spelling and grammar errors as you type

Figure 10-1 shows how the spelling and grammar checker highlights errors as you type. Here, the spelling errors are underlined in red and the grammar errors are underlined in blue. Then, you can correct the errors as you go, or you can page through the document after you've finished it and correct the errors.

In the example in this figure, the first highlighted error is a grammar error because "addition" should be followed by a comma. The second one is a spelling error. When the corrections are obvious, you can just correct the errors, and that will remove the underlining.

Otherwise, you can right-click on the error to display a shortcut menu like the one in this figure. Then, you can select the suggested correction. In this example, "addition" will be replaced by "addition" with a comma after it.

For grammar errors, you can also click on Ignore Once to ignore the correction so it won't be highlighted in the document. You would do that if the highlighted error is actually correct. Or, you can click on Grammar to get more information about the error.

For spelling errors, you can click on Ignore Once or Ignore All in the shortcut menu to ignore the word if the spelling is correct. You can click on Add to Dictionary to add the word to the dictionary so it will no longer be incorrect. Or, you can click on Add to AutoCorrect so that misspelling will be automatically corrected in the future.

A document with spelling and grammar errors

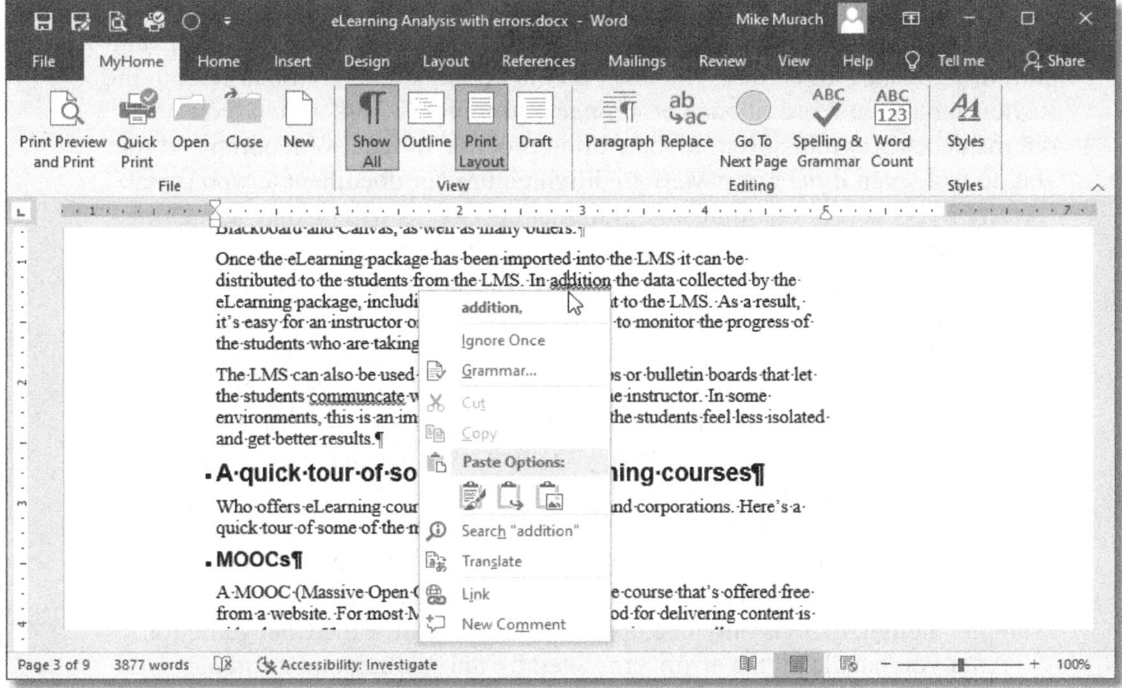

How to correct grammar errors

- If you right-click on a grammar error, the shortcut menu may include one or more correction possibilities. If you click on one of them, it replaces the highlighted words.
- The shortcut menu for a grammar error may also include these commands: Ignore Once and Grammar. If you click on Grammar, the Grammar pane is opened as shown in the next figure.

How to correct spelling errors

- If you right-click on a spelling error, the shortcut menu may include one or more correct spelling options. If you click on one of them, it replaces the highlighted word.
- The shortcut menu for a spelling error also includes these commands: Ignore All, Add to Dictionary, and Add to AutoCorrect.

Description

- By default, spelling errors are underlined in red, and grammar errors are underlined in blue…as you type.
- If you recognize the cause of a highlighted error, you can fix it right away. Otherwise, you can use its shortcut menu to get correction options and more information.

Figure 10-1 How to fix spelling and grammar errors as you type

How to run the spelling and grammar checker

Another way to catch spelling and grammar errors is to run the spelling and grammar checker. To do that, you click on the Spelling and Grammar icon in the Review tab of the Word ribbon, as summarized in figure 10-2. Often, you will run the checker as a final step in the editing process. In fact, we recommend that you do that, even if the errors were highlighted in your document as you typed.

After you start the spelling and grammar checker, it will display a Spelling or a Grammar pane for each error that it detects. Then, you can decide whether you want to fix, ignore, or skip each error. If your document has many errors, of course, this can be a time-consuming process.

In the first example in this figure, you can see how the checker can help you correct a grammar error. Here, you can see the highlighted error in the document, and you can see that the error is described in more detail in the Grammar pane. Then, if you agree with the correction, you click the Change button. Otherwise, you can click Ignore Once.

In the second example, you can see how the checker can help you correct a spelling error. If there is more than one correction option in the Spelling pane, you select the correction that you want and click the Change button. In this example, though, there is only one option so you just click the Change button.

After you handle all the errors presented by the spelling and grammar checker, the readability statistics are displayed as shown in the next figure. Then, if you run the checker again, just the statistics are displayed. The errors aren't checked again.

So if you do want to recheck the errors for the document, you have to open the dialog box in figure 10-4 and click on the Recheck Document button. Then, you can rerun the spelling and grammar checker.

As you use the spelling and grammar checker, you should keep in mind that it won't catch all errors. If, for example, you enter *type* when you mean *typo*, the error probably won't be caught. Similarly, the grammar checker won't catch many of the grammatical errors that are common in all writing. That's why proofing is a critical step in the development process, whether or not you use the spelling and grammar checker.

A grammar error caught by the spelling and grammar checker

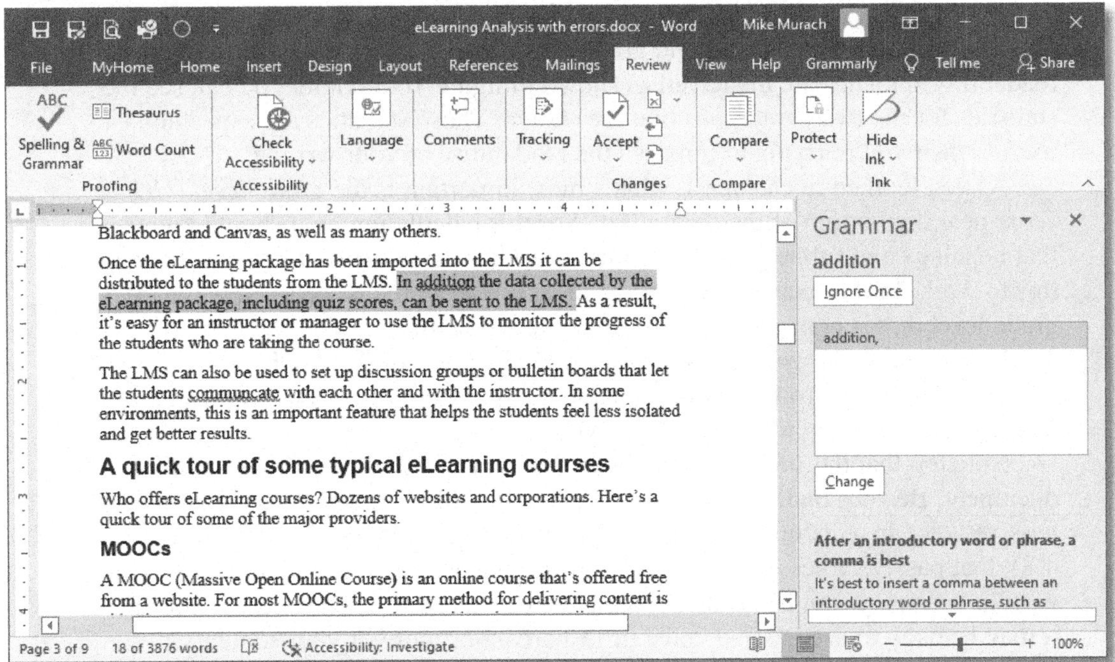

A spelling error caught by the spelling and grammar checker

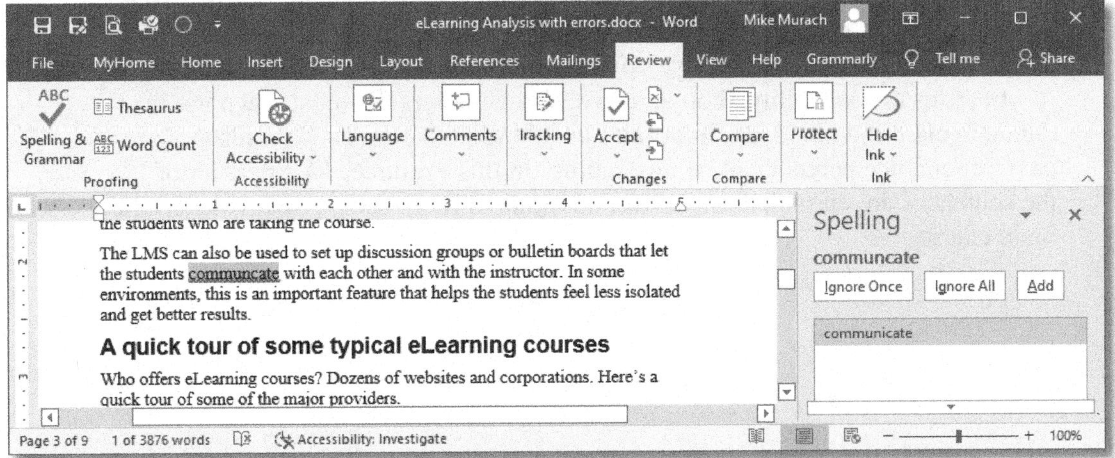

How to use the spelling and grammar checker

- To run the checker, press F7 or click the icon in the Review tab of the ribbon.
- When a spelling or grammar error is detected, a Spelling or a Grammar pane is opened. Then, you can select from the options in that pane. After you respond to all of the suggested changes, the readability statistics are displayed as shown in the next figure.
- To check for errors again, you need to click Recheck Document in the dialog box shown in figure 10-4, and then run the spelling and grammar checker.

Figure 10-2 How to run the spelling and grammar checker

How to get the readability statistics

After you run the spelling and grammar checker and fix the errors, the readability statistics are displayed as shown in figure 10-3. Here, you can see the statistics for the two paragraphs that are shown. These statistics are especially useful when you're trying to improve the readability of your writing.

As you learned in chapter 4, what's most important is the *grade level* (*GL*) *score* near the bottom of the dialog box. For this calculation, you should know that headings and subheadings aren't included, which is the way you would want this to work. In this example, the grade level is 8.3, which is well below the 10.5 grade level that is reasonable for most business documents.

If you follow the guidelines presented in chapter 4 of this book, you shouldn't have any problem with readability. But if you're getting scores of 12 or more, you really need to improve your writing.

Note too that the statistics include the percent of passive sentences in a document. Besides that, the grammar checker highlights many of the uses of passive voice in a document as errors. But remember from chapter 4 (figure 4-8), that passive voice is frequently used in business, technical, and educational writing because it can be used to reverse the emphasis of a sentence. As a result, it may be okay to use passive voice in 15 percent or more of your sentences.

You also should realize that the grammar checker counts a sentence as passive whether the passive voice is in the main clause of the sentence or in an adjective, noun, or adverbial clause. For instance, the last sentence in the first paragraph in this example is counted as passive, even though active voice is used in the main clause and the passive voice is in the concluding adverbial clause.

In terms of readability, though, active voice matters the most when it's in the main clause, and it's less important in subordinate clauses. As a result, the passive sentence percent can be misleading. In this example, 14.2 percent of the sentences are classified as passive, but none of them use passive voice in the main clause.

Two paragraphs from this chapter with one passive sentence

Another way to catch spelling and grammar errors is to run the spelling and grammar checker. To do that, you can click on the Spelling and Grammar icon in the Review tab of the Word ribbon, as summarized in figure 10-2. Often, you will run the checker as a final step in the editing process. In fact, we recommend that you do that, even if the errors have been highlighted in your document as you typed.

After you start the spelling and grammar checker, it will display a Spelling or a Grammar pane for each error that it detects. Then, you can decide whether you want to fix, ignore, or skip each error. If your document has many errors, of course, this can be a time-consuming process.

The readability statistics for the two paragraphs

Readability Statistics ? ×

Readability Statistics	
Counts	
Words	127
Characters	556
Paragraphs	2
Sentences	7
Averages	
Sentences per Paragraph	3.5
Words per Sentence	18.1
Characters per Word	4.2
Readability	
Flesch Reading Ease	67.8
Flesch-Kincaid Grade Level	8.3
Passive Sentences	14.2%

OK

About the statistics

- The calculation of the *grade level (GL) score* doesn't include headings, although they are included in the word, character, and paragraph counts.
- A sentence is counted as passive if the passive voice is used in any clause of the sentence, not just the main clause.

Description

- By default, the readability statistics are displayed at the end of a spelling and grammar check. If they aren't, you can check the Show Readability Statistics box in the Options dialog box that's shown in the next figure.
- If you make some changes and want to get the statistics again, you can run the spelling and grammar checker again. This gives you the statistics without redoing the spelling and grammar check.

Figure 10-3 How to get the readability statistics

How to set the spelling and grammar options

If you're happy with the default options for the spelling and grammar checker, you don't need to change them. However, you should at least be aware of what they are. You may also want to turn some options off if they aren't finding errors that need to be corrected.

The basic spelling and grammar options

Figure 10-4 shows you the default settings for the basic spelling and grammar options. In the third group of options, you can see that the spelling and grammar checker will check both spelling and grammar as you type. It will also show readability statistics after it finishes checking the spelling and grammar.

If you decide that you don't like to see the highlighted errors as you're working, you can uncheck the Check Spelling as You Type and the Mark Grammar Errors as You Type options. In fact, you may want to work that way when you write the first draft of a document. That way, you won't get bogged down with details as you concentrate on writing the paragraphs that deliver your ideas. Then, as the last step in the editing process, you can run the spelling and grammar checker.

In general, the default options in this dialog box work the way you want them to, unless you want to turn off the check-as-you-type options. However, if you click on the Settings button that's in the third group, the dialog box in the next figure is displayed. In that dialog box, you may want to change some of the default options.

After you run the spelling and grammar checker and handle all the errors, the readability statistics are displayed. Then, if you make some changes to the document and run the checker again, the readability statistics are recalculated and displayed but the errors aren't checked. To change that, you can click the Recheck Document button in the Word Options dialog box before you rerun the checker.

The Word Options dialog box

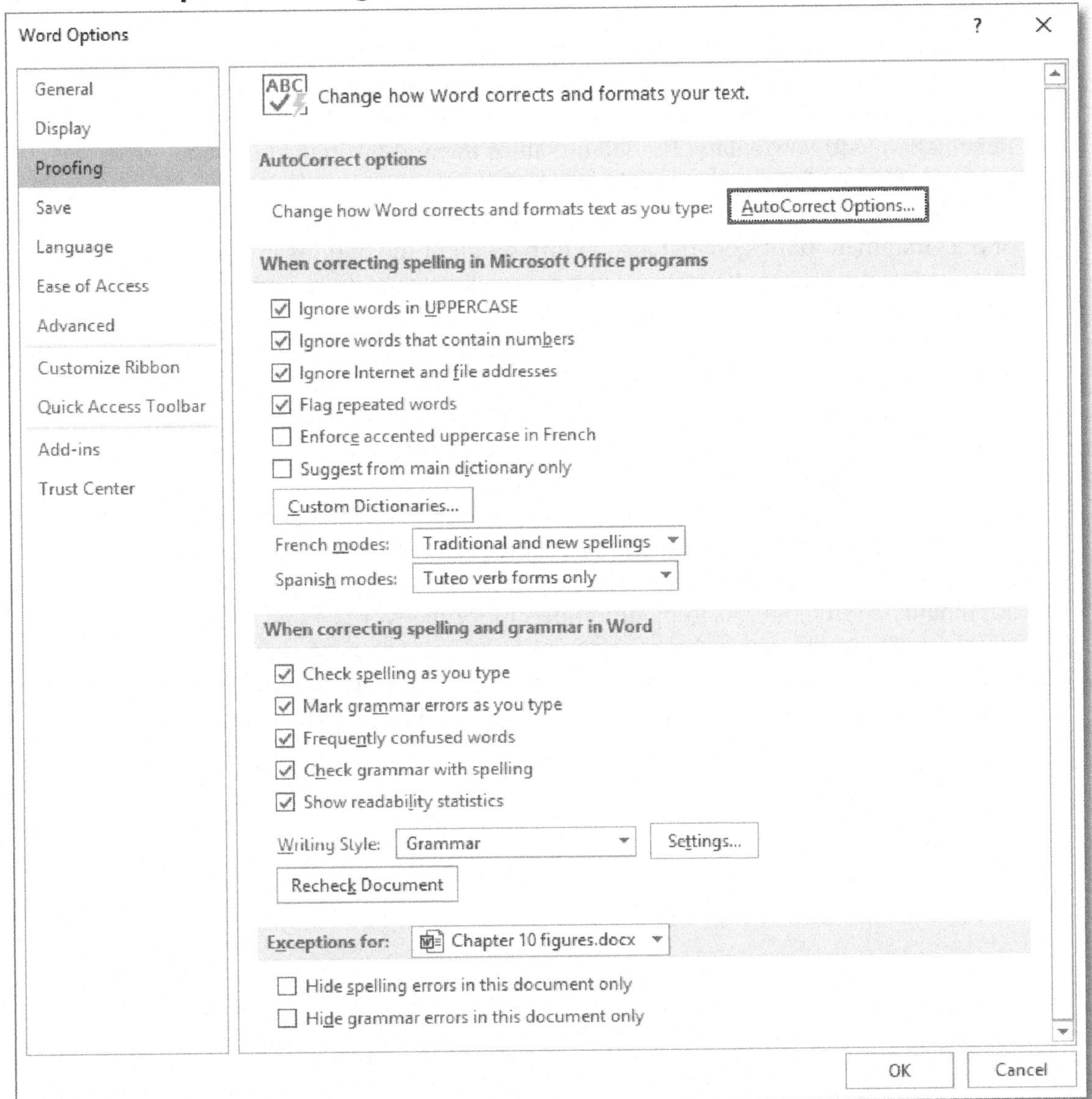

Description

- To access the spelling and grammar options, use the File→Options command and click on Proofing in the left pane.
- If you set the options as shown above, both spelling and grammar will be checked as you type.
- If you check Show Readability Statistics, the statistics are displayed at the end of a spelling and grammar check.
- If you click on the Settings button, the options in the next figure are displayed.

Figure 10-4 The basic spelling and grammar options

Other grammar options

Figure 10-5 shows the Grammar Settings dialog box that you access from the Proofing tab of the Options dialog box. This dialog box provides options in these groups: Grammar, Clarity, Conciseness, Formality, Inclusiveness, Punctuation, and Vocabulary. By default, all of the options in the Grammar group are on, but none of the options in the other groups are on.

When you start using the spelling and grammar checker, it's good to keep all of the Grammar options on, but also to turn on all of the options in the Clarity and Conciseness group. You may also want to check the Clichés option in the Vocabulary group to see whether it will help you identify trite language.

Then, as you get experience with these options, you may decide that you don't like the results that you get from some of them. For instance, you may want to turn off the Passive Voice option because it marks too many instances that are okay the way they are. That way, you'll have fewer suggestions to review when you run the spelling and grammar checker.

Of course, to set the options the way you want them, you need to understand what they refer to. For instance, Comma Splice refers to two sentences that are separated by commas instead of periods, which is something you shouldn't do. Indefinite Article refers to the improper use of the words *a* and *an*. And Nominalizations are words that are derived from other parts of speech by adding a suffix to them. That option helps to identify errors that are in keeping with the chapter 4 recommendation to avoid the use of words that have prefixes and suffixes.

The Grammar Settings dialog box with all options in the first three groups turned on

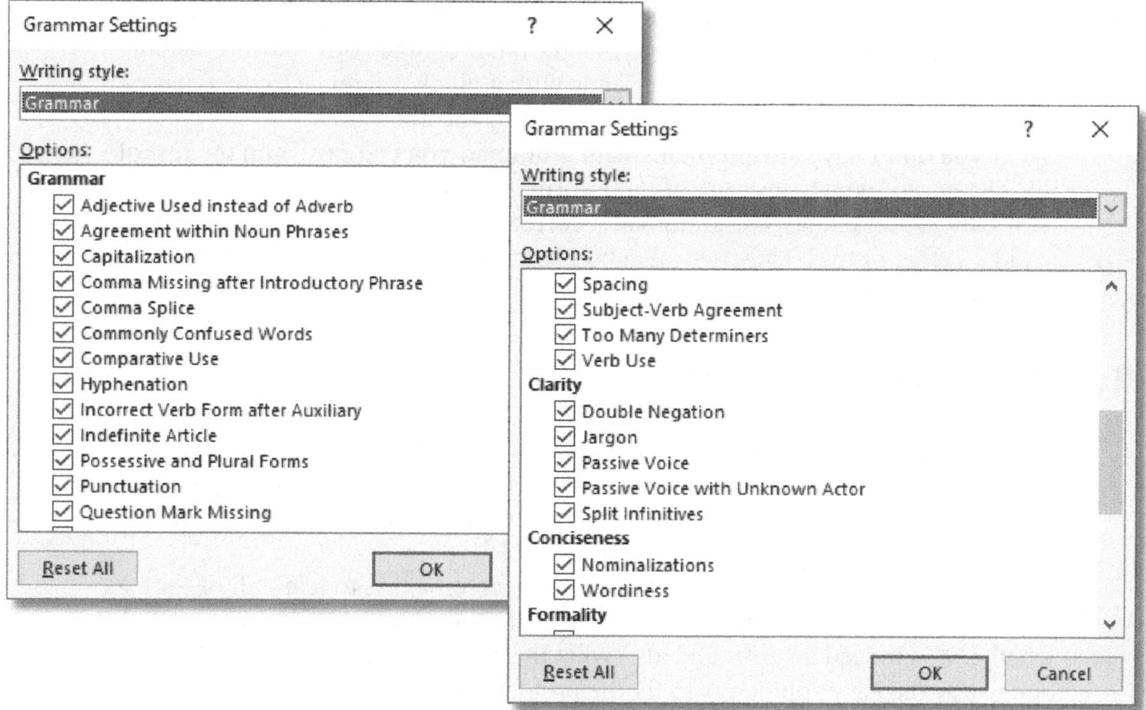

Description

- To display the Grammar Settings dialog box, click on the Settings button in the Options dialog box that's shown in the previous figure.
- The Grammar Settings dialog box has groupings for Grammar, Clarity, Conciseness, Formality, Inclusiveness, Punctuation Conventions, and Vocabulary.
- By default, all options in the Grammar group are on, but all of the options in the other groups are off.
- To start, you should turn on all of the options in the Clarity and Conciseness groups to see whether they will be useful. Then, after you've used the grammar checker for a while, you can turn off the options that aren't helpful.
- When you run the Spelling and Grammar Checker, you'll find that some errors aren't really errors. You'll also find that the checker doesn't find all of the errors.

Figure 10-5 Other grammar options

A quick reference for grammar

Because the spelling and grammar checker identifies some errors that aren't really errors, you need to understand the grammar behind some of the checking that's done. To that end, this chapter ends with a quick reference to the types of grammar errors that are likely to be identified by the checker.

If you don't have any problems with grammar, you can just skim the rest of this chapter to see whether any of the information is of use to you. But if you aren't always sure that your grammar is correct, these summaries may help you avoid an error or two. They may also help you identify areas in your skillset that you need to improve.

Subject-verb agreement

Figure 10-6 presents some typical problems that business writers have with subject and verb agreement. The trouble is that the spelling and grammar checker doesn't always detect this type of error. And if it does detect an error, it may not offer the right correction.

So, the best approach to subject and verb agreement is to learn all of the rules in this figure. Then, if the checker identifies an error, you'll be able to tell whether it is indeed an error and also what the best correction is. If you're not quite sure, the easy solution is to fix the problem by rewriting the sentence.

Make sure the verb agrees with the subject

Wrong: *One* of every five new books *are* successful.

Right: *One* of every five new books *is* successful.

Compound subjects connected by *and* take plural verbs

Wrong: Our *plant* on the north end of Chicago *and* our *factory* on the river in Ohio still *produces* all our products.

Right: Our *plant* on the north end of Chicago *and* our *factory* on the river in Ohio still *produce* all our products.

Take the number of a compound subject connected by *or* or *nor* from the second-named noun or pronoun

Wrong: Either Jake or *Tom have* to make the trip.

Right: Either Jake or *Tom has* to make the trip.

Treat most collective nouns as singular

Wrong: Your entire *staff have been invited* to the meeting.

Right: Your entire *staff has been invited* to the meeting.

With a collective noun, use a plural verb if it describes the actions of the group as a unit

Wrong: The *majority* of the students polled *doesn't* want a dress code.

Right: The *majority* of the students polled *don't* want a dress code.

Treat company names as singular unless they sound as if they're plural

Right: System Associates, Inc. *has* its own staff of writers.

Right: The System Associates have their own staff of writers.

Treat indefinite pronouns such as *anybody* and *everybody* as singular

Wrong: *Everybody* leaving the seminar sessions *need* passes.

Right: *Everybody* leaving the seminar sessions *needs* a pass.

For a fraction, make the verb agree with the noun that follows

Wrong: Two-thirds of the *rent are* due today.

Right: Two-thirds of the *rent is* due today.

Description

- When the spelling and grammar checker detects an error in subject-verb agreement, it doesn't always offer the right correction.
- When in doubt, it's best to rewrite the sentence so you're sure that the agreement is written correctly.

Figure 10-6 Subject-verb agreement

Pronoun references, dangling modifiers, and misplaced words and phrases

Figure 10-7 presents some typical problems that business writers have with pronouns, modifiers, and misplaced words and phrases. Here again, the spelling and grammar checker may not catch this type of error. And if it does, it may not suggest the correct change.

The first two groups of examples demonstrate two types of errors that are relatively common in business documents. First, you need to make sure that the *antecedent* for a pronoun is clear. In other words, the reader must be able to tell what the pronoun refers to. Second, you need to make sure that a pronoun and its antecedent are both either singular or plural.

The third group of examples shows how to fix a modifier when what the modifier refers to isn't clear. This is known as a *dangling modifier*. The trick is to identify the error when the grammar checker doesn't. Once it's identified, you can rewrite the sentence so the modifier isn't dangling.

The last two groups in this figure present a similar problem. That is to make sure that it's easy to tell what your adverbs and prepositional phrases refer to. These are problems that the grammar checker can't detect, so the burden is on you to find and correct them when you edit a document. The usual solution is just to move the modifier closer to what it's modifying.

Unclear pronoun references

Wrong: After the managers and staff members met, *they* felt thwarted.

Right: After *their* meeting with management, the staff *members* felt thwarted.

Incorrect number agreement between pronouns and antecedents

Wrong: *Each* of the managers brought *their* portfolios.

Right: *Each* of the managers brought *her* portfolio.

Wrong: Either Doug or Steve should be responsible for the plan, and they should have full power to implement it as they see fit.

Right: Either *Doug* or *Steve* should be responsible for the plan, and *he* should have full power to implement it as *he* sees fit.

Dangling modifiers

Wrong: *Without adequate training,* this system can't solve your productivity problems.

Right: Unless your staff has adequate training, this system can't solve your productivity problems.

Wrong: *Based upon your needs,* we're sending the pricing information for the equipment we discussed.

Right: Here is the pricing information for the equipment that we agreed is based on your needs.

Misplaced adverbs

Wrong: He answered all the questions asked by the staff members *patiently.*

Right: He *patiently* answered all the questions asked by the staff members.

Misplaced prepositional phrases

Wrong: I am looking for a job with salary and benefits that are appropriate for my education and experience *with room for professional growth.*

Right: I am looking for a job that provides fair compensation and an opportunity for professional growth.

Description

- When you use a pronoun, you need to make sure that its *antecedent* is clear and that it agrees in number with that antecedent.
- You also need to make sure it's easy to tell what your modifiers, adverbs, and prepositional phrases refer to.
- A *dangling modifier* occurs when a sentence doesn't make clear what the modifier refers to.

Figure 10-7 Pronoun references, dangling modifiers, and misplaced words and phrases

Comma use

Figure 10-8 presents most, if not all, of the comma rules that you should use for business writing. Today, business writers often ignore comma rules like these, and yet all of these rules will make your documents easier to read. That's because commas help show the structure of your sentences.

For instance, the first rule in the first group in this figure is to use commas after introductory words, phrases, and clauses. Those commas show the reader where the main clause of the sentence begins. For instance, in the third example for this first rule, a comma is used after an introductory word (*Then*) as well as after an introductory adverbial clause.

The second rule in the first group is to separate *all* of the items in a series. That includes putting a comma after the last item before the word *and* or *or*. Although that last comma is often omitted in business writing, using it will help prevent misreading.

The third rule in the first group is to set off words that interrupt the thought of the sentence. And the fourth rule is to use a comma before a concluding adverbial clause that changes the flow of the sentence. Note, however, that a comma isn't needed before a concluding adverbial clause that doesn't change the flow of the sentence.

The second group of examples in this figure presents some standard comma rules that you should obey. Remember that a *non-restrictive phrase* or *clause* is one that isn't essential to the meaning of the sentence.

One other type of comma fault that isn't shown in this figure is using a comma where one isn't needed, which is likely to confuse the reader. To avoid that problem, you should just use commas when the rules in this figure call for them, and not use them anywhere else.

Comma use for improved readability

After introductory words, phrases, and adverbial clauses

To start, you run the Spelling and Grammar Checker.

When you run the Spelling and Grammar Checker, it displays a dialog box for each error or misspelling that it detects.

Then, when you respond to a dialog box, the checker looks for the next error or misspelling.

To separate *all* items in a series

The primary eLearning components are video, slides, simulations, and quizzes.

To present that skill, you can use video, slides, or simulations.

To set off words that interrupt the thought of a sentence

Beyond that, however, eLearning packages are used for corporate training.

If, for example, you're getting complaints, they should be your primary concern.

Before a concluding adverbial clause that changes the flow of the sentence

He couldn't have done a better job in the interview, although his skillset doesn't seem to fit our job.

Standard comma use

To set off non-restrictive phrases and clauses

Mr. Clark, who is about 35 years old, made a good impression.

To set off contrasting phrases

He selected the standard, not the innovative, approach.

To separate independent clauses in a compound sentence

Tom conducted the interview, and he made the selection.

To separate two or more adjectives that each modify the noun

A shy, quiet, soft-spoken person would not be right for this job.

To set off the years within dates

He arrived on August 31, 1988, and stayed for years.

To separate the parts of an address or location

She lives at 2019 Erie Street, Racine, Wisconsin.

Description

- Although comma use in business writing is less rigorous than it once was, all of the uses in this figure will help make a document easier to read.

Figure 10-8 Comma use

Capitalization and italics

In case you have any trouble with the use of capitalization or italics, figure 10-9 summarizes their common uses. For capitalization, the most frequent uses are for *proper nouns*, names derived from proper nouns, points of the compass when they refer to parts of the country, and the titles of books, magazines, and articles. If you don't know whether a word or group of words should be capitalized, you can look it up on the internet.

Except for the uses in this figure, the trend is to use fewer capital letters in business writing. For instance, you should only capitalize the first letter of headings and subheadings, not the first letter of all the main words in them. You should also use lowercase to refer to visuals, as in figure 1, not Figure 1. By using fewer capital letters, you improve readability because most people can read lowercase letters more easily than capital letters.

This figure also shows the primary uses for italics in business writing. The first two are standard. The third one is commonly used in business and technical writing to introduce new terms or key terms.

Capitalization use

Proper nouns

Pepsi Cola	Queen Victoria	Waldorf Hotel
St. Agnes Hospital	Monday	September
Times Square	Maple Street	Yellowstone Park
Mary Jackson	Revolutionary War	Magna Carta
Old Testament	the Almighty	Australia

Names derived from proper nouns

an American	British hospitality
Spanish history	a Virginian

Points of the compass when they refer to parts of the country

the West	the Northeast

In the titles of books, magazines, and articles

The Art of Readable Writing

The Wall Street Journal

The article called "The Art of Delegation"

Italics use

For the titles of books, newspapers, magazines, and journals

The Art of Readable Writing

The Wall Street Journal

Business Week

For letters used as letters and words used as words

Don't pronounce the *t* in the word *often*.

Business people like to use a language of *however*s and *therefore*s.

To introduce a new term or a key term

We call this document a *heading plan*.

Description

- You probably don't have any trouble with the use of capitalization and italics, but this figure provides a quick reference if you need one.
- The trend today is to use less capitalization because that makes your writing easier to read. That's why you should only capitalize the first letter in report headings and subheadings, and you should avoid other unnecessary capitalization.

Figure 10-9 Capitalization and italics

Misspelled and misused words

The most glaring errors in business documents are often misspellings. If you misspell a word in an email to your boss or a prospective employer, the misspelling is liable to be noticed before anything else. When you use Word, however, the spelling checker should identify any misspellings. Nevertheless, misspellings of the words in the first part of figure 10-10 still slip into business documents.

Even more glaring and more common is the misuse of a word like one of those in the second part of this figure. In fact, it's easier to forgive a misspelling because that can be just a clerical error. In contrast, the misuse of a word shows that your vocabulary is limited. Nevertheless, misused words are also relatively common in business documents.

So, ask yourself whether you know the difference between the words in each pair in this figure. Do you know, for example, that *reticent* means unwilling to reveal one's thoughts and that *reluctant* means unwilling to do something? Do you know that *farther* should be used when you're talking about distances and *further* should be used when you're talking about movement in an abstract sense? That's the type of word precision that will improve your writing.

If you do have trouble with words like these, you need to be sure that you know their precise meanings before you use them. The good news is that it's easy to get the information that you need by searching the internet. That's why there's no excuse for misusing a word today.

Frequently misspelled words

absence	accommodate	acquainted	amateur
benefited	bureau	cafeteria	cemetery
coarse	conquer	deceive	desirable
desperate	eighth	exhausted	fascinating
February	guarantee	guardian	irresistible
maintenance	manageable	misspelled	necessarily
noticeable	permissible	personnel	persuade
possibility	preceding	preferred	prejudiced
privilege	repetition	rhythm	sergeant
sincerely	stationery	succeed	their
there	transferred	villain	weird

Frequently confused or misused words

about/around	adapt/adopt
affect/effect	allusion/illusion
amount/number	complement/compliment
continual/continuous	disinterested/uninterested
famous/notorious	farther/further
fewer/less	good/well
ingenious/ingenuous	liable/likely
persecute/prosecute	principal/principle
reluctant/reticent	stationary/stationery

Description

- The Spelling and Grammar Checker should help you correct any misspellings. But the Spelling and Grammar Checker may not be able to help you correct misused words or technical terms that aren't in its dictionary.
- When in doubt about the meaning of a word, you can right-click on it and select Synonyms from the shortcut menu to get the synonyms for it. Or, you can select Search to get links to definitions and related information on the internet.

Figure 10-10 Misspelled and misused words

Perspective

It's always worth taking the time to check and fix the errors that are caught by the spelling and grammar checker. That's why you should run the checker as the last step in the editing process. Another benefit of running the grammar checker is that it displays readability statistics after you fix the errors.

If you have any trouble understanding the suggestions that the spelling and grammar checker makes, you probably need to improve your grammar skills. That's also true if you have trouble understanding the grammar summary that's presented in this chapter. In either case, you can search the internet to get the help that you need.

Remember too that a spelling and grammar checker can't begin to find all of the errors in a document. So even if it helps you correct an error or two, you're the only one who can make sure that your documents are correct. That's why proofing is always an essential step in the writing process.

Terms

spelling and grammar checker
grade level (GL) score
antecedent
dangling modifier
non-restrictive phrase
non-restrictive clause
proper noun

Summary

- The *spelling and grammar checker* helps you find and fix the errors in a document. You can see the errors as you type, or you can run the spelling and grammar checker after you finish writing a document.
- When the spelling and grammar checker finishes checking a document, it displays the readability statistics for the document including the *GL score*.
- Because the spelling and grammar checker doesn't always suggest the right corrections, you still need to understand the rules of grammar.
- To be an effective business writer, you need to make sure that your subjects agree with your verbs, that your pronouns have clear *antecedents*, that your pronouns agree in number with their antecedents, and that you don't have *dangling modifiers*, misplaced adverbs, or misplaced prepositional phrases.
- To be an effective business writer, you should use commas to improve the readability of your sentences. You should also use the standard rules for capitalization and italics, but otherwise minimize your use of capitalization.
- To avoid embarrassing errors like misspellings and misused words, you need to use the spelling checker to catch the misspellings. You also need to use the internet to look up the differences between words that are frequently confused.

Appendix A

How to download and install the reports and templates for this book

If you want to review and experiment with the reports and templates that are illustrated in this book, this appendix shows how to download and install them. That includes Word and RTF (Rich Text Format) versions of two of the eLearning reports and the two Word report templates that are presented in chapter 8.

How to download and install the reports and templates for this book

Figure A-1 shows how to download and install the files for this book. In brief, go to our website, go to the download page for this book, and download the zip file to your computer. Then, double-click on the zip file to extract the files into a folder named tested_methods that contains the subfolders shown in the first table in this figure. At that point, you can move the tested_methods folder wherever you want it.

The second table in this figure gives the names of the two reports that are downloaded. Then, you can use the Word versions if you use Microsoft Word as your word processor. Or, you can use the RTF versions if you're using a different word processor.

The third table in this figure gives the names of the two Word templates that are described in chapter 8. If you're using Word, you can use these templates to start new documents and then apply the styles in the template to the paragraphs in the new documents.

The Murach website

www.murach.com

The folder for the downloaded files

\tested_methods

The subfolders for the downloaded files

Folder	Description
reports_rtf	Rich Text Format (RTF) versions of two of the eLearning reports
reports_word	Word versions of two of the eLearning reports
templates	Word versions of the two Report templates presented in chapter 8

The names of the reports

Filename	Description
Why most eLearning fails	A short report with one heading level
Is eLearning an opportunity	A long report with two heading levels

The names of the templates

Filename	Description
Report	A report template with these styles: Report title, Heading 1, Heading 2, Normal, Indented, Bulleted, and Numbered
Report with TOC	The Report template with a table of contents on the first page.

How to download and install the files for this book

1. Go to www.murach.com.
2. Find the page for *Murach's Tested Writing Methods.*
3. Scroll down to the "FREE Downloads" tab and click it.
4. Click on the DOWNLOAD NOW button for Reports and Templates. That should download a zip file named tested_methods.zip.
5. Double-click on the downloaded zip file to extract the reports and templates into a folder named tested_methods.
6. Use File Explorer (Windows) or Finder (macOS) to move the tested_methods folder to wherever you want it.

Figure A-1 How to download and install the reports and templates for this book

Appendix B

One of the reports that illustrates the tested writing methods

This appendix presents one of the two reports that are used throughout this book to illustrate the tested writing methods. This report illustrates the use of headings, the principles of paragraphing, and the art of writing readable sentences. As this book recommends, this report was drafted just once, edited once, and proofread once. It wasn't written, edited, rewritten, edited, and edited again with the goal of making it perfect.

As appendix A shows, you can download this report as well as the longer eLearning report that's used to illustrate the use of this book's methods. The downloaded reports are available in both Word and RTF (Rich Text) formats.

Why most eLearning fails

During the last several years, eLearning has taken on a much larger role in both corporate training and college instruction. And it all seems so simple. The trainees or students can take the courses anywhere, anytime, and on any type of device, whenever they have a few minutes. And the trainers and instructors can use their Learning Management Systems to monitor the results of their courses.

But shouldn't someone be asking whether these courses deliver the results that they promise? And even if they do deliver the results, shouldn't someone be asking whether eLearning is the most efficient way to deliver those results? Because by my analysis, most of these courses don't deliver the results that they promise, and even if they do, they are painstakingly slow.

What follows, then, are some of the reasons why most eLearning is both ineffective and inefficient...or simply put, why most eLearning courses fail. These reasons apply to eLearning courses for many skillsets, but especially for skillsets that are complicated. In fact, the more complicated a skillset is, the less likely it is that eLearning will be an effective way to teach it. Or, to say that another way, eLearning seems to work best for trivial skills.

The limitations of video

Video is one of the primary mediums for delivering the content of eLearning courses. And there's no question that video is effective whenever you can learn more by seeing how something is done than by reading about it. For instance, video is effective for showing how to dance, how to play tennis, and even how to use an IDE (Integrated Development Environment) for a programming course.

But for most subjects, video has some serious limitations. That is,

- Most people can read from 3 to 5 times faster than a video presents information. So even if a video is effective, it's an inefficient way to learn.
- It's hard to skip over the video information that you already know because you can't be sure where it ends. That slows you down even more.
- It's hard to refer back to the information that you need to review. But that's essential for any complicated subject.

To illustrate, suppose you watch a video of a computer procedure, something as simple as how to use Excel formulas. When you finish it and try to do the steps that you just watched, it's hard to remember what they were. It's also hard to do the steps at the same time that you're watching the video because you have to switch back and forth between the video and the software product that you're trying to master.

The limitations of slides

For many years, PowerPoint slides were the primary medium for business presentations as well as for classroom instruction. In either case, the presenter describes and enhances the information that's on the slides. In recent years, slides have also become a primary medium for delivering the content in eLearning courses. The trouble is that slides have some serious limitations when it comes to presenting information.

The primary limitation is that slides force you to break down the content into chunks that are so small that it's hard to see how the chunks relate to each other. But it's seeing those relationships that lead to deeper understanding and the ability to apply what you've learned. For programming subjects, this means that there's no way to present something like a lengthy program listing without breaking it down into pieces that don't make sense by themselves.

Like video, the other limitation of slides is that it's hard to skip over information that you already know and it's hard to review information. Although both are easier to do with slides than with video, it's still time consuming. And that's compounded by the fact that the information has been broken down into chunks.

What most eLearning courses don't include

Because of the limitations of video and slides, most eLearning courses are missing some critical components. That includes:

- Examples that are complex enough to illustrate how the skills are used in the real world
- Practice exercises (not quizzes) that ask the students to apply what they've learned...and solutions that let them learn from their mistakes
- Reference materials that help the students review what they've learned and apply those skills on the job

Although it's true that some eLearning courses do include these components, the reference materials are usually minimal so they can't begin to make up for the limitations of the videos and slides.

Why the focus on technology is counterproductive

I think the root cause of why most eLearning courses fail is that the authoring tools force the course developers to focus on how they're going to deliver the content instead of the best way to deliver it. That focus is not only on how to use video and slides, but also on what types of quiz questions to use, when and how to use simulations, what kind of interactions and transitions to use, whether to add gamification (and awards)...and recently, when and how to use Virtual Reality.

The trouble is that none of those features will contribute much to an effective course. In fact, the more the technology takes the student away from the skills that the course is supposed to teach, the less effective the course is likely to be. If, for example, the goal of a course is to teach Python programming, what the student should be doing most of the time is using a Python IDE to code, test, and debug Python programs. Anything that takes away from that...like fancy quiz questions or games that deliver certificates...reduces the effectiveness of the course.

A related point is that once the course is over most of the students will need to enhance their skills. To do that, however, they usually won't be able to watch a video or take a slideshow or play a game. Instead, they'll have to read books, technical manuals, or website pages. But that's one more reason why the focus on technology is counterproductive. In the long run, the skill that the students are going to need the most is how to get the information that they need by *reading. So that's what they should be doing as they're learning.*

My conclusions

eLearning will always have its advocates because it makes the job of the instructor so easy. Just set the courses up and evaluate the results. And in truth, eLearning can be effective for short courses with limited learning objectives.

But for subjects like the ones that our books present, eLearning is both ineffective and inefficient. That doesn't mean that there aren't eLearning courses for the same subjects that our books present. But any fair evaluation of learner outcomes will show that our books are far more effective than the eLearning courses. In fact, the evaluation will probably show that the eLearning courses fail.

Index

G

H

I

J

L

M

N

O

P

Q

R

S

T

V

W

XYZ

Why most business writing fails

- Poor methods or no methods for planning what you're going to write (see chapters 1 and 2)
- Ineffective use of headings and subheadings (see chapters 1 and 2)
- Poor paragraphing (see chapter 3)
- Failure to show the relationships between paragraphs and between sentences (see chapter 3)
- Sentences that are hard to read and understand (see chapter 4)
- Using visuals in a way that doesn't make a document easier to read and understand (see chapter 5)
- Editing while you're writing instead of getting the first draft on paper as quickly as possible (see chapter 6)
- Editing, rewriting, and editing again, which we call "thrashing" (see chapter 6)
- Failure to use the word processing features for writers: templates, styles, the outline feature, and the spelling and grammar checker (see chapters 8, 9, and 10)

The downloadable reports and templates

- The eLearning reports that are used as examples in this book, in both RTF and Word formats, so you can review the complete documents
- The Report and Report with TOC templates that are presented in chapter 8 so you can start new Word documents from them

How to download the reports and templates

- Go to murach.com, and go to the page for ***Murach's Tested Writing Methods.***
- Scroll down the page until you see the "FREE downloads" tab. Then, click on it and proceed from there.
- For more information, please see appendix A.